TECHNOLOGY TRANSFER:

Financing and Commercializing The High Tech Product Or Service: From Research To Roll Out

KARL J. DAKIN
JENNIFER LINDSEY

PROBUS PUBLISHING COMPANY
Chicago, Illinois

© 1991, Karl J. Dakin and Jennifer Lindsey

ALL RIGHTS RESERVED. No part of this publication may be reproduced, stored in a retrieval system, or transmitted by any means, electronic, mechanical, photocopying, recording, or otherwise, without the prior written permission of the publisher and the copyright holder.

This publication is designed to provide accurate and authoritative information in regard to the subject matter covered. It is sold with the understanding that the publisher is not engaged in rendering legal, accounting or other professional service.

Library of Congress Cataloging in Publication Data Available

ISBN 1-55738-160-7

Printed in the United States of America
BC
1 2 3 4 5 6 7 8 9 0

Dedication

To Barbara Walder and Darla Dakin, each for their unique support

Contents

Acknowledgements   ix

Chapter 1   **An Entrepreneurial Perspective of Technology Transfer**   1
Technology and Its Transfer   2
Reality and Myth   4
Technology Transfer Players   9
Risks and Rewards   15
Trends   19

Chapter 2   **The Technology Life Cycle**   21
The Cycle Model   23
Calendar and Timelines   38
Points of Transfer   39

Chapter 3   **The Technology Transfer Process**   41
Seller Identification of Technology: Step One   42
Buyer Identification of Technology: Step One   46
Seller Identification of Buyer Market/Industry: Step Two—Part I   48

	Buyer Identification of the Inventor and Seller: Step Two—Part II 49
	Seller Validation of Technology: Step Three—Part I 50
	Buyer Validation of Technology: Step Three—Part II 51
	The Sale of the Technology: Step Four—Part I 53
	The Buyer's Deal: Step Four—Part II 60
	Seller Transfer Technology: Step Five—Part I 62
	Buyer Transfer Technology: Step Five—Part II 63
Chapter 4	**Technology Transfer Mechanisms 65**
	Less Than Complete Transfers 66
	Straight Sale 67
	Publication 69
	License 70
	Resale 71
	Cooperative R&D 72
	Contract R&D 73
	Loaned Servant 74
	Merger 74
	Spin-Off 76
	Reverse Engineering 77
	Theft 79
Chapter 5	**Planning/Management Skills 81**
	Technology Transfer as an Objective 83
	Technology Transfer as a Strategy 89
	Resources 91
	Planning 95
	Tactics 97
Chapter 6	**Marketing Skills 99**
	Benefits/Needs 99
	Identification of Technology End-Users 101
	Identification of Prospective Buyers 104
	Packaging 107

Incentives 107
The Presentation 108
The Technology Program in Oak Ridge 109
When Finding Buyers Is Difficult 120
When Qualifying Buyers Is Difficult 122
Identification of Sellers 122

Chapter 7 **Financial Skills** **125**
Government Financing 125
Tech Transfer Costs 130
Accounting Considerations 132
Seller Financing 132
Buyer Financing 133
Types of Financing 133
Sources of Financing 138
Price of Money 140
Financing and the Deal Structure 140
Forms of Compensation 143

Chapter 8 **Legal Skills** **145**
Intellectual Property 145
Licensing 147
Nondisclosure Agreements 148
Contracts 149
Fraud and Misrepresentation 149
Business Structures and Relationships 150
Government Regulations 151
International 151
Dispute Resolution 153

Chapter 9 **International Transfer** **155**
International Transfer Protection 156
Global Licensing 156
Global Patenting 158
International Trade Secrets 158
International Trademarks 159
International Copyright 160

    Export Restrictions 160
    Currency/Foreign Exchange Exposure 161
    Cultures and Customs 162

Appendix A Resources for Technology Transfer 165

Appendix B Legislation 173

Appendix C Sample Agreements 175

Index 231

Acknowledgements

This book reflects a contribution towards a comprehensive, yet simple methodology, upon which the science of technology transfer may advance. We would like to thank the following individuals who assisted in some way by contributing ideas, motivation, materials and comments: Kathleen Byington, Colorado State University Research Foundation; Dick Chapman, Chapman Research; Lou Higgs, Center for the New West; Syl Houston, U.S. Western Executive Seminar Center; David Leavitt, U.S. Small Business Administration; Dana Moran, Solar Energy Research Institute; Tim Peach, Teletronics; and Jerry Plunkett.

The human mind is often so awkward and ill-regulated in the career of invention that it is first diffident, and then despises itself. For it appears at first incredible that any such discovery should be made, and when it has been made, it appears incredible that it should so long have escaped men's research. All which affords good reason for the hope that a vast mass of inventions yet remains, which may be deducted not only from the investigation of new modes of operation, but also from transferring, comparing, and applying those already known, by the methods of what we have termed "literate experience."

Francis Bacon
Novum Organum, Book I

1
An Entrepreneurial Perspective Of Technology Transfer

Technology transfer should be a strategy, not a last resort.

An illuminating way to view technology transfer is as an athlete who anticipates the Olympics knowing he is only one player among the thousands who will compete in the giant global game. An entrepreneur should perceive his role in technology transfer in much the same way. Like the Olympics, technology transfer is an ancient pursuit and a performance benchmark achieved successfully—for the most part—only by world-class players.

The technology Olympics in the 1990s, however, will be even more competitive because there will be many more players. In the past, large corporations had the limelight and made the profits. In this decade, there will be exponentially more competitive, entrepreneurial athletes willing to try for the gold, from among sole proprietorships, small businesses, educational institutions and government agencies. Although the rules of the game remain intact for technology transfer, these new competitors have broadened the scope of the game, so to speak, and continue to call for new rules of play.

To understand how small companies with annual revenues of under $25 million can hope to compete against world-class competition, this chapter describes technology transfer from the perspective of the small business owner as an entrepreneur who is a technology seller or rightsholder. (These terms will be used interchangeably throughout the text.) Chapter One helps the entrepreneur get in the game by explaining the first crucial step he will have to take to become an adept player: recognizing and accepting that acknowledged expertise comes only from experience in the technology transfer game itself. In fact, no other corporate or financial expertise seems to apply.

To explain the calls, this chapter includes a game plan that helps entrepreneurs distinguish between the reality of global transfer and the fantasies that most hopeful technology sellers harbor about their creative ideas and how to commercialize them. The idea is to strip the technology transfer process of the glamour and the promise of instant wealth, and lay out a workable commercialization strategy that can be implemented when the time is right for the inventor, the technology and the marketplace.

Technology and Its Transfer

Technology most commonly is perceived as the ground between science and business, but a ground which rapidly can turn into a financially-debilitating abyss if these disciplines are not merged by the technology rightsholder who also can function as a sophisticated entrepreneur. Technology actually can bridge these two very demanding masters when it is commercialized successfully—even when science and business are head-to-head in competition instead of working in harness for cooperative goals.

When applied to technology transfer, the exacting disciplines of science aim to increase knowledge that represent universal truths as we perceive them. But knowledge in and of itself has minimal value in the marketplace if consumers cannot do something useful with the knowledge, like eat it or drive it to the supermarket. So the benefit of knowledge only emerges when it is converted to a usable form.

The disciplines of the corporate world, on the other hand, increase the value of knowledge in the form of profits, which are a reward for the assumption of risk and the commitment of resources in supplying products and services to the world as usable knowledge. Without the new knowledge provided by science, the value of existing products and services would remain static.

In technology transfer parlance, then, technology is the conversion

Chapter One

or commercialization of new knowledge into enhanced products and services. The ownership of technology, therefore, carries with it a significant responsibility to complete the commercialization of knowledge in exchange for the opportunity to be rewarded generously for those risks and commitments. Historically, it has been the anticipation of those rewards that motivates entrepreneurs and inventors to assume the high risks and financial burdens of technology advancement through commercialization.

But the cost of technology commercialization is very often burdensome. During the transition from knowledge to product, technology undergoes evolutionary change until it becomes a product or service. The seven formal steps required to commercialize knowledge can be performed by one individual or by mega-sized laboratories staffed by hundreds of scientists. And the process of shifting the responsibility for performing any of the seven major commercialization tasks is what is known as technology transfer.

Following is an illustration of the role of knowledge in the technology transfer process.

A Four-Stage Process of Technical Advance:

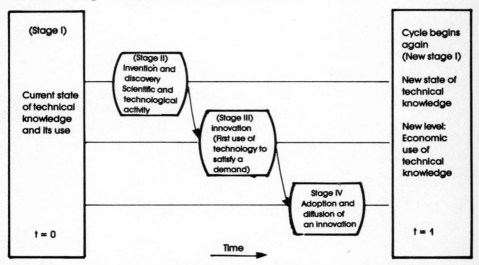

Source: *Factors in the Transfer of Technology*, William H. Gruber and Donald G. Marquis

Reality and Myth

Although the term "technology transfer" was coined only about 20 years ago, technology transfer's reputation has become legendary. As an amalgamation of corporate fantasies, technology transfer has a downside as well as an upside: it has destroyed many of the most prevalent myths of American business culture and it has helped support the foundation of the American corporate profit system. The most successful entrepreneurs in the technology transfer process are capable of distinguishing clearly between the fact and the fantasy inherent in this age-old trading practice.

One of the most common fantasies is based on a well-known American myth about inventions. The premise always has been accepted—among the unknowing—that the world beats a path to the door of the entrepreneur who builds a better product, creates an invention or makes a "breakthrough" discovery. This myth has been decimated by marketing gurus in recent years amid growing global competition that demands a credible presentation of the product or service before any sale will be executed in the increasingly selective marketplace—no matter how innovative the product is.

This proof also applies to technology before it has been completed. One consequence of incomplete technology—a technology that has not been processed through the seven steps described in Chapter Two—is the need for additional marketing over and above the marketing required for a finished product.

Another common myth about technology is the wisdom of "do-it-yourself" commercialization. There is an unquenchable desire—or need—among many entrepreneurs to wear all the hats as inventor, financier, business manager and salesperson. Their attitude often is, "who knows the business/product/technology better than I do?" But the entrepreneur who hopes to complete technology commercialization all alone usually fails to acknowledge his limitations in terms of talent, resources, energy, know-how and sales ability.

In fact, it is rare for one individual or organization to complete the commercialization process successfully without the considerable benefit of outside expertise. So one of the most obvious values in technology transfer is shared responsibility for commercialization;

"R&D Success and Its Predicators"

- Technical Success Does Not Equal Commercial Success
- Success Is A Function of Environmental Opportunity (Technological and Economic), Internal Organizational Design and External Linkage Structures
- Small Firms vs. Large Firms

Source: The Industrial Technology Institute

The most difficult myth to shatter is that small, entrepreneurial companies can achieve commercial success by meeting, or even exceeding, technical standards without the array of resources available to the large corporation: capital, management, technical personnel and important strategic alliances that help the company get technology to market.

---

this allows everyone in the development and transfer pipeline to utilize his skill specialization most effectively and profitably.

Although there is a place for everyone in the technology transfer process, it is far from a panacea for all sales, research and development, or profit problems that arise in American business; in some instances, technology transfer can actually be detrimental to the success of technological development.

A third myth says technology transfer is a straight sale: a near-instantaneous exchange in which the technology goes to one party, and money and other considerations go to the other party. The reality of an actual transfer, particularly a global transfer, belies the seeming simplicity of the transfer process in a most dangerous way.

For example, technology transfer in recent years has been touted as the magic antidote to overwhelming foreign competition as well as the stimulus for significant local economic development projects. As later chapters in this book show, this misconception about technology transfer and a little snake oil will not cure all that ails American business.

Ideally, technology transfer should be treated as an alternative strategy for accomplishing the sales, revenue or profit goals of a singular business—not as an economic strategy for large geographic, economic or political entities. It would be fair to say that the usefulness and potential benefit of technology transfer increase directly in proportion to how well the internal goals of its corporate parent are planned and executed.

A fourth myth is that the technology subject to transfer should be limited to or comprised only of the initial concept or primary technology. In order for any technology to be commercialized successfully, it requires the creation of two additional, or secondary, technologies. Failure to develop these technologies can portend the ultimate failure of the primary technology's success in the marketplace.

The first additional technology provides a way to manufacture the original or primary technology in quantities of scale sufficient to meet market demand. For example, few users would buy a computer chip if it had to be hand-crafted and cost $1 million per chip. The price of the original technology at the user level is of primary importance in determining the net benefit received by the user.

The second additional technology provides education and assistance to the user who must implement the original technology. Few users would buy a computer in which the chip arrived without a user's manual, technical diagrams or some form of instructional material. This type of technology is prevalent in the user-friendly characteristics of successful primary technology. It is an attempt to make technology benefits match directly with the needs of the user.

The creation of these secondary technologies is essential to commercialize successfully; in fact, full commercialization of the primary technology requires incorporation of the secondary technologies as though they represent integral parts of the primary technology.

Technology in Evolution

The primary objective of technology transfer is the successful shift of responsibility for one or more of the seven tasks which are required to commercialize a technology in the marketplace in final form. The key phrase here is "final form," because the primary obstacle to successful transfer is the perpetual state of change which the technology must

Chapter One

undergo before and usually after the technology reaches the marketplace. Starting as a concept, technology undergoes a lengthy series of changes until, eventually, it can be utilized by the end-user. This progression or evolution is referred to as the "technology life cycle," and is described in detail in Chapter Two.

The technology transfer process would be simplified greatly if this on-going evolution did not occur during the commercialization process. But the very essence of profit is change; so as science increases the world's storehouse of knowledge for conversion into technology, this evolution enhances nearly all products and services destined for the marketplace.

The Rules of Technology Transfer

Technology transfer is one game for which there is no recognized set of rules for the players. If a rulebook could be compiled, of necessity it would have to be a text massive enough to encompass every known complexity in the world of global trade.

This section of the chapter, therefore, defines some basic rules for the technology transfer competition: it addresses the entrepreneurial perspective; it provides an overview that synthesizes the common game plans without getting into the nuances of expert play or the options that are unique to an individual transfer. This chapter contains the steps that should be taken to complete only a generic technology transfer.

In reviewing these generic technology transfer steps, however, it is important to note that the primary objective of any technology development is seldom the transfer itself. A successful technology transfer is nearly always a secondary objective or byproduct of another activity which supports the primary objective of the entrepreneur. In fact, the secondary or derivative nature of most transfers accounts for the confusion that can complicate the transfer process: the ultimate transfer often looks like an accident or simple good fortune for an industrious technology seller. So to isolate the elements of the process that indicate success more easily, this book examines technology transfer as though it were the primary objective instead of the secondary objective, and then it positions transfer within the context of the principal objectives for all of the players.

Technology Transfer System (Participants)

Sender

- Socio-economic environment of the reference company

- Reference company, sender or lessor of the industrial know-how
  - Skills / Other company activities
  - Existing reference company skills

- Company(ies) Subcontractor(s)
  - Other company skills
  - Existing technical skills

- Companies joining in a new industrial project
  - The engineering consultancy
  - Project leader
  - Constructor
  - Builders

- Lessors

- Labor supply companies

Receiver

- Socio-economic environment of the receiver company

- Receiving or acquiring company
  - Technical competences to be created
  - Other company activities

- Company(ies) Subcontractor(s)
  - Technical competences to be created or improved
  - Other company activities

- Builders
- Constructors
- Subcontractors of foreign contractors

Technology Transfer System

Technology Transfer Consultant

Source: Technology Transfer: A Realistic Approach, Silvere Seurat, Gulf Publishing Company

This diagram shows how the participants in a technology transfer are linked to the transfer system.

Chapter One

Technology Transfer Alternatives

Technology developers can choose from among many strategies to accomplish their corporate aims, including the transfer of responsibility for completing the commercialization process of a technology. Each strategy for winning—however success is viewed by the entrepreneur—involves unique benefits and detriments; therefore, some transfer strategies are inferior or superior to others.

An important caveat for all entrepreneurs who consider technology transfer is to review similar mechanisms for bringing their technologies to the marketplace, contingent on costs, the time required for development and approvals, the viability of testing, the local economy and many other considerations.

Required Skills

A successful transfer demands that an entrepreneur practice a multitude of skills in order to bring the technology to market, including specialized abilities in management, marketing, finance, law, accounting and international trade, among others. Although these are considered to be requisite skills in the efficient operation of most small businesses—especially those in which the owner wears all the hats—these common business skills take on new dimensions of significance in the operation of a company that plans a technology transfer.

Technology Transfer Players

In light of the diverse ways in which technology transfer is accomplished in the 1990s, several well-defined sectors of the corporate world are considered major players in the transfer process. Each of the following sectors are comprised of individuals as well as large populations who may have nothing more in common than the objectives of the group:

- Corporations, including small businesses
- Government agencies and departments
- Educational facilities
- Facilitators in the public and private sectors

Because corporations are the deep-pocket research and development players, they are the largest slice of the technology transfer pie. Precise statistics have not been compiled to date on the dollar volume of annual research and development in the United States, but current estimates total between $60 billion and $120 billion in 1989 alone, depending on the size companies included in the calculation for U.S. businesses. These expenditures include the dollars spent on new and/or improved technology and on technology transfer.

As will be demonstrated, technology transfer is big business for the corporation that is aggressive about profitability. But the caveat is that technology transfer can help or hurt profit objectives because a new product can bring increased sales or it can obsolesce existing products that require extensive re-tooling in order to make the company profitable again. Because technology is a focal point for competition, technology transfer presents a method for remaining or becoming competitive.

Although many businesses are virtually unaffected by technological change, other firms fade into Chapter 11 when competitive new technologies are introduced into the marketplace. So technology can be the pivot of competition among individual companies, as well as among entire industries. The impact of technology and its transfer is contingent on the company's industry, its competitive edge, its depth of resources, and its ability to look down the pipeline and read the future accurately.

"Top Transfer Performers"

- Top 100 Universities, 80 percent of Academic R&D
- Top Tier of Non-Mission Federal Labs
- 213,000 Individuals (out of 1.2 million)
- 15 percent Women, 2 percent Blacks (universities)

Source: The Industrial Technology Institute

Chapter One

One of the most significant factors in the success of a transfer is the level of variance in resources between large and small companies. Because most large corporations are well aware that technology transfer is a complicated transaction, they assemble large development teams who specialize in a multitude of disciplines to perform the five steps of technology transfer (see Chapter 3). Other large companies that elect not to undertake the complexities of transfer refuse either to acquire or to sell technology, adopting the "NIH" attitude: not invented here. This is the attitude that prevails among entrepreneurs as well.

The key to the transfer decision—whether or not to attempt a transfer—is the potential impact of a transfer on the company. Technology transfer can be most vital even when it is the last available strategy by which to avoid failure. But the last-ditch effort usually occurs when the entrepreneur has exhausted the corporate coffers before the technology has reached the marketplace. Technology transfer in this instance can occur even when failure means a complete shutdown of the business and the technology has been shelved in order to pursue other business ventures.

Generally, when sales are lagging or when the company faces bankruptcy, an entrepreneur at least entertains the fantasy that a white knight is waiting in the wings to rescue the operation by replenishing the checking account. But more realistic entrepreneurs are learning to create technology transfers that allow them to complete the commercialization of the technology, recapture initial investments and possibly even earn a profit.

For many entrepreneurs, technology transfer is a superior strategy for achieving corporate goals when they give up the idea of completing the commercialization of the technology without outside help. These technology sellers shift the responsibilities for completion at the point in the life cycle when their profits can be maximized.

They begin to limit corporate risk by retaining only those development activities they enjoy or do well. This divestiture of responsibility, cost and risk creates efficiencies that lower the overall cost of getting the technology to market and improve the technology's potential success.

If the technology seller is not driven to a transfer because of declining operations or bankruptcy, the transfer process holds great opportunity—much like panning for gold—in which a lucky strike can yield legendary wealth. Other even more savvy entrepreneurs scour the business world for under-valued and under-utilized technologies that have been conceived by other scientists and inventors, creating a powerful convergence with enough resources to maximize all of the technologies' benefits through the efforts of many developers.

Government

Federal, state and local governments also play a strong role in the conception and development of new technology. In order to enhance the health, safety and welfare of its citizens, for example, governmental departments and agencies both initiate and finance research and development projects even when they are considered too risky or unprofitable in the private sector. When government initiates or supports the technology development process, ultimate transfer usually is not considered the primary objective.

It is important to note that only on occasion is technology transfer designated the primary objective of the contracted project so that a business or educational facility can reap the rewards that result from their research and development efforts in partnership with the government.

Federal government R&D spending in 1989 totaled approximately $60 billion dollars, according to government reports, with the majority earmarked for defense contractors and educational institutions. Until just a few years ago, the government's principal technology transfer "strategy" simply was to give research and development results away. Under the theory that the research was financed with tax dollars, the government generously conceded to return the resulting technology to the taxpayers.

Of course this giveaway was criticized by the more profit-minded members of Congress and the bureaucracy; eventually this criticism resulted in the government's current practice of implementing precise and profitable technology transfer mechanisms that accrue to the government's "bottom line."

Royalties Collected By Agencies: 1986–1988

Agency	Amount Collected
Agricultural Research Service	$ 213,416
NIST	104,312
NOAA	11,492
Air Force	57,244
Army	28,535
Navy	20,048
Energy/Fossil Energy[a]	0
EPA	0
Bureau of Mines[b]	54,000
USGS	0
NASA[c]	181,760
NIH[c]	3,946,263
Total	$4,617,070

[a]The zero is for the Department of Energy's Office of Fossil Energy, which is responsible for the government-operated energy technology centers. Other entities within the Department of Energy collected royalties totaling about $881,000 for this period, but this amount was not subject to distribution under the 1986 act because the inventions were made by contractors.

[b]Bureau of Mines estimate.

[c]The amount shown for NASA is for the period October 1986 through December 1988.

[c]All of NIH's royalties were for agreements made prior to the 1986 act and were primarily for the National Cancer Institute's acquired immune deficiency syndrome-related inventions.

Source: U.S. General Accounting Office

Local economic developers—banks and other financial institutions in partnership with municipal governments—often view technology transfer as prospecting. Because economic development experts are charged with leveraging the area's existing resources in order to create a wider tax base or more jobs, they all too often invest in big deals that purportedly will recoup the costs of past failures or declining economies. It is a tall order, but economic development can benefit from the prudent use of technology transfer when transfer promotes economic efficiency in general and improves the success rate of local entrepreneurs on an individual basis.

Technology transfer has another tall order in the national scheme of things for the 1990s. The private and public sectors now consider transfer a strategy for maximizing its last and most valuable natural resource: technology. It is well-known, and now anticipated, that exporting technology actually can improve the balance of foreign trade because in the long run, costly giveaways of base technology may be more harmful to the economic health of the nation than the trade imbalances of the 1980s.

Education

Educational institutions also have a significant interest in technology transfer because their primary goal is to transfer knowledge. It is important to note that, historically, transferring knowledge has been synonymous with discovering new knowledge and they often are perceived as one process. But in light of technology transfer, the manner in which new knowledge is processed still reflects education's original objective to transfer knowledge.

Although technology transfer has been increasingly important to the fund-raising efforts and reputations of educational institutions, university-based transfers generally are not structured to make a profit: typically, educational institutions, like the government, have created knowledge and then given it away as an unrealistic return of taxpayer dollars.

In 1980, the federal government gave a further boost to research conducted by educational institutions. The Bayh-Dole Act of 1980 gave small businesses and nonprofit organizations the right to claim title to any technology created during performance of a research

---

"Characteristics of Successful University-Industry Linkage"

- Hybrid Organizations
- Historically Stable, Multilevel Relationships
- Land Grant/Technical Orientation, Not Private Elite
- Flexibility on Intellectual Property

Source: The Industrial Technology Institute

---

contract with a federal agency. This allowed a large volume of federal research to be claimed by educational institutions.

Facilitators

The complexities inherent in the technology transfer mechanisms in place for the 1990s have lead to the creation of a new professional population that has emerged expressly to make the transfer process more efficient and profitable worldwide. Usually referred to as "facilitators", these individual and corporate consulting firm experts play an increasingly significant role in making technology transfer happen.

When a market has not yet been developed for the purchase and sale of a technology, facilitators support the commercialization process rather than become directly involved in the transfer. The forms of support they offer include providing information, planning conferences, staffing expositions, and helping or creating technical organizations, among other functions. Unlike corporations, governments and educational facilities for whom transfer is a secondary objective, facilitators view transfer as the primary objective.

Risks and Rewards

Technology commercialization presents many risks to the entrepreneur. But the foremost risk is the entrepreneurial investment in time, money, facilities, energy and human resources in a technology. If the entrepreneur's technology fails the economic proof, it usually is a great loss.

The Unknown

The technology life cycle consistently asks two challenging questions of the entrepreneur: (1) is the technology scientifically valid? (2) is the technology economically valid? Positive responses to these questions are the "proofs" which technology commercialization requires. These proofs are necessary, although demanding: if the technology fails either proof for any reason—say noncompliance with government regulation, or an incorrect premise—the entrepreneur probably will not succeed in the commercialization process. To proceed down the development pipeline without these positive proofs is gambling dangerously with money, time and the status of the company.

The first proof addresses the primary concern of the government, and of the people and institutions that finance entrepreneurial research: does it work? What they want to know more about is what the technology can deliver in terms of true user benefits. At this point in the process, usually early on, the question is asked in the abstract without regard to the economic proof which must be considered separately at a later point in the process.

The second proof, the economic proof, concerns the payback to any entity involved with the commercialization of the technology. It answers the question, "will the investment in commercialization of the technology be rewarded with a profit?" Initially, this test should be applied to the life cycle of the technology as a whole. Then the test should be applied to each stage, or to each task within a stage, of the life cycle. Because the risk of performing each task varies, it may be possible for an entrepreneur to perform one task and achieve a profit—even if the technology ultimately fails commercially.

Commitment of Resources

Because of the accelerating pace of technology development, opportunity is measured in units of time in the development and transfer process. That means entrepreneurs should view the process as a window that offers a limited amount of time in which to champion one or more technologies. Betting on the wrong technology usually results in an irreplaceable loss of opportunity that is hard to recoup, especially when competitors bet on the right one.

In fact, the criteria for judging a successful technology is fairly hardnosed in the world of global transfers: even when a profit/loss analysis for a given technology indicates a green light, a successful technology is viewed as a loss in the big picture when newer technologies with more potential are developed at the same time.

Capitalization also plays a key role in technology commercialization, so standard financial tools should be used to measure the validity of technology transfer. For example, each stage in the technology life cycle should be subjected to a separate return on investment (ROI) analysis. In this way, the entrepreneur can pick and choose among the various tasks of commercialization in order to make the smartest capital investment relative to his resources. This selective assumption of tasks encourages the entrepreneur to specialize, which leads to his improved efficiency; this, in turn, leads to greater profits because the entrepreneur can transfer away those tasks that are low-yielding or that are better performed by others.

The mental and physical energy required to commercialize the technology also must be factored into the commitment equation. Some products take only a few weeks or months to develop, either because the owner has been working on the concept for a long period of time prior to the onset of the development process, or because the technology itself is so simple that it requires little mental exertion.

But most technologies require a significant outlay of mental and physical energy, in a momentum that must be sustained by the entrepreneur in order to complete the life cycle stages of development in a timely and competitive way.

When any number of development crises strike—lack of financing, a hitch in the original hypothesis, among others—he must be prepared to sustain a burst of physical energy that will carry the technology through to completion despite the setbacks.

Payoff

Although the rewards of transfer occasionally are as lucrative as entrepreneurs hope, more often they do not come close to matching the risks that have been assumed in order to commercialize the technology. Payoffs can range from complete failure, in which the total investment was lost, to profits that return far less than conservative bank or so-called "blue chip" investments.

"Economic Impact of Innovations"

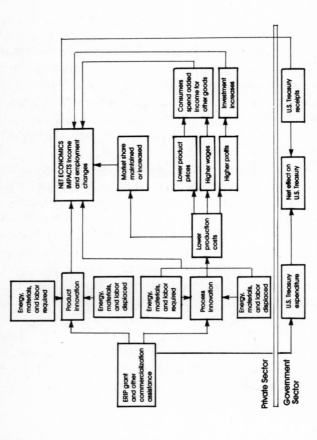

Source: "New Emphases: Technology Transfer Sandia National Laboratories," Robert P. Stromberg, Technology Transfer Institute, 12th Annual Meeting, p. 288

Included in even a casual study by the potential transferor of the economic impact of technology transfer should be a review of how commercialization stimulates new demand for energy, materials, equipment and labor—which also generate secondary employment benefits.

The technology transfer payoff for technology developers can range from as little as one-quarter of one percent—or lower—in consumer products with low proprietary interest (little or no novelty to the product and one subject to high competition) to as high as 60 percent returns in software products. Most royalty rates range between 1 and 5 percent.

Trends

Because of the enormous risks and commitments required to bring technology to market, transfer is here to stay as a viable corporate strategy. As more and more dollars are earmarked for research and development, there will be more new technologies on which to build enhanced products and services for the future needs of the global marketplace.

At the same time, there will continue to be an important stratification between new knowledge and its use, requiring more and greater specialization in the performance of the tasks necessary to get a new technology to market, as well as a resulting increase in the number of technology transfers.

In addition, the enactment of recent government legislation (see Appendix B) will spur the migration of technology from federal laboratories to the private sector. Supported by a similar migration from educational laboratories, the pure marketplace for new technologies will be stimulated to even greater growth.

2
The Technology Life Cycle

The life cycle formula can be applied to virtually 100 percent of technologies brought to the marketplace; omitting any step in the life cycle—especially the proof stage—substantially increases the possibility of commercialization failure.

One of the fundamental characteristics of technology is most critical to understanding technology transfer: technology development is in a constant state of change until it reaches final form, and sometimes it continues changing even after it reaches the marketplace. Any technology begins with a concept and undergoes a myriad of transformations until its ultimate use by an end-user.

This continual change is a double-edged sword: it can mean a miraculous evolution of products and services for the world; but change also means that the act of completing a transfer, in itself, is made more difficult.

At the onset because change is never completed, it is difficult—if not impossible—to project with any accuracy to what end change will deliver the technology. The potential results of this level of inaccuracy can be a technology that does not work, or a technology that is not profitable even if it does work.

This dilemma, of course, creates substantial risk for the developer and can make technology transfer a potentially undesirable transaction in some instances for the developer as entrepreneur. In addition, technology usually increases in value during, or as a consequence of,

Technical Advance Related to Demand-Technical Feasibility Fusion

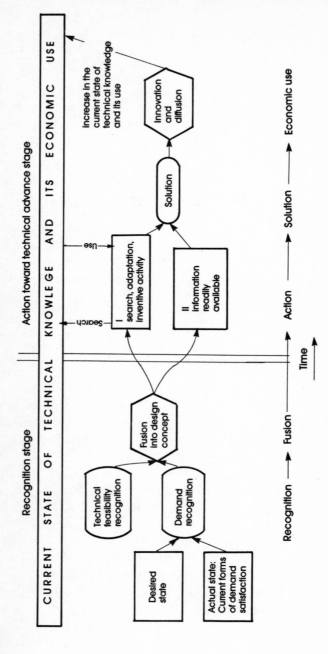

Source: Factors in the Transfer of Technology, edited by William H. Gruber and Donald G. Marquis, The M.I.T. Press

Innovation is most likely to be initiated when recognition of market demand and a workable technical solution for the demand are integrated, and when feasible technical response to user requirements occurs.

the transfer. This makes the assessment of price even more difficult.

In the same way, a critical characteristic of technology transfer is that typically it takes place before the technology reaches final form. This means that the technology at the time of transfer does not possess all of the characteristics that it will have in its final, usable form. In essence, the transfer takes place while the technology is still in the process of evolution. This also poses other distinct difficulties for the parties who attempt to complete a transfer.

The Cycle Model

To better understand the impact of the fundamental characteristics of technology and its transfer, it is necessary to illustrate the evaluation of a technology within a life cycle—or the stages through which technology must pass in order to reach a commercial form suitable for marketplace sale:

Concept
Research
Development
Manufacture
Distribution
Sale
Use

An understanding of these life cycle stages can help entrepreneurs as technology sellers identify the benchmarks most technologies must achieve in order to be commercialized, and the activities which must

Utilization of Space

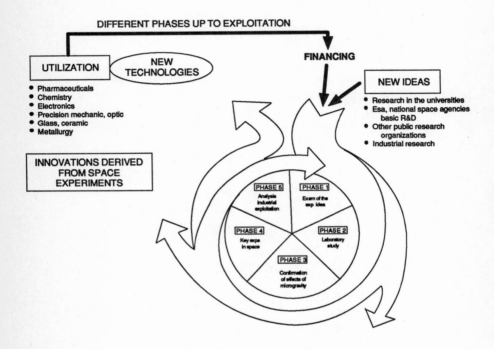

Source: Technology Transfer and the Development of a Commercial Space Industry, Jeffrey L. Struthers, Center for Space and Advanced Technology

This commercialization and transfer model was derived from government/private sector research efforts operant within four parameters: limited availability of capital, institutional and political uncertainty, high transportation costs and high degree of risk.

be completed to produce a scientifically and economically successful technology. It is important to note here that only about 2 percent of all technologies patented result in a commercial product; of these, only half are considered successful in the marketplace. Although it is possible and even necessary to break the life cycle of development into even more stages for certain types of specialized technologies, the steps described below apply to most technologies.

Within each stage, the entrepreneurial technology developer must assess the technology in terms of:

- Its importance
- The objectives to be completed
- Its characteristics
- Potential risks and rewards
- The opportunity to transfer

Concept

The starting point in the life cycle of any technology is a concept, the most important characteristic of which is its intangible nature. As outlined in Chapter 1, this initial concept as the first step toward technology transfer must represent new scientific knowledge that is perceived to be a step forward from prior knowledge.

This step forward can be an insignificant development or it can be a big leap in developmental knowledge. In either case, this advance of scientific knowledge is a natural progression which is cumulative from past information; its existence allows other scientists and the corporate world to forecast even further based on this data as an incremental improvement over existing knowledge.

Other advances in scientific knowledge are so crucial that they are perceived as a "breakthrough." Generally, a scientific breakthrough is not considered predictable in any way; and, unlike an incremental move forward in knowledge, a breakthrough can be conceived by a single scientist or by a large staff of developers.

It is not uncommon for many researchers to make the same discovery within a very short period of time. The technological concept can be novel, the only one of its kind or the first of its kind.

The key to a breakthrough is that the concept is unique, thereby enhancing its market value.

At some point during the concept stage of the technology, the developer of the concept—or the rightsholder—must take the initial step toward commercialization by activating the technology's life cycle. This usually takes the form of (1) disclosing the concept to another individual, (2) writing down the concept on paper, or (3) research and development testing. All three of these options have the same result: it is conceded by the marketplace that taking the first step changes the characteristics of the technology.

In fact, most technologies are viewed as more valuable after the first step is taken than before that step is taken. This important change creates an impact both on the ownership of the technology and on its perceived value.

Most developers of a technological concept, at this point, try to estimate the value of their concept. When the phrase "excellent idea" or "brainstorm" is used to describe the idea, they hope it conveys the ability to measure potential value immediately.

This may be true of some technologies when the concept is so simple, so capable of immediate understanding and appreciation, and so ready for immediate application, that the marketplace is willing to attribute an arbitrary value to the concept before it progresses further into the life cycle. But in general, it is more common for a technological concept to need some major work before a market value can be assigned.

So it is important for technology sellers to remember that, 99 percent of the time, the technology must progress to the next stage of its life cycle before it can be valued; and until it does, the technology remains only a concept to the marketplace.

In fact, a large percentage of the marketplace says technology in concept stage has no value, despite the potential enhancement and profit it represents. The reason is that most technologies require so much additional development that the costs of that development can outweigh the ultimate value of the concept.

If the technology does have a future in terms of enhancement and eventual profit, the concept stage is crucial for another reason: it represents the point in time when ownership attribution is determined in order to establish appropriate forms of legal protection and for publication as the designated discoverer or inventor of the technology.

Although discovery and ownership are important early on, technology transfer itself is virtually impossible at this point. Not only has there been no scientific or economic proof established to satisfy market valuation, the enormous costs of commercialization remain undetermined.

And the technology developer still has not matched the risks and costs with a realistic assessment of potential profit or enhancement. So the payment of any significant amount of money for a technology in concept stage is hardly possible before the technology advances further along the development continuum, or life cycle.

To move toward eventual technology transfer, the entrepreneur should first go through a planning activity similar to the one described in Chapter 5. This practice allows him to size up the technology in terms of risks and rewards, available resources and his capability of completing the steps for eventual commercialization.

The planning practice also helps the entrepreneur identify that point at which it will be most profitable to transfer the technology to a third party.

Some technology sellers actually do transfer conceptual technologies without due preparation for—or actual advancement toward—the next stage in the technology's life cycle. But a transfer at this point precludes thousands to millions of dollars in profits and licensing fees that could have accrued had they gestated the technologies longer and sold a more mature product.

Research

At the concept stage, a technology is no more than an assumption about new knowledge. It is considered only a perceived truth which is untested and unproven. In fact, the marketplace says it is no more than a hypothesis, a "what if" that may be true or false when subjected to professional testing.

So the objective of the research stage is to test the scientific principles on which the concept is based. The research stage of the life cycle ends conclusively when the developer can state, without reservation, that the hypothesis has proven correct, that the concept now

The Total Innovation Process

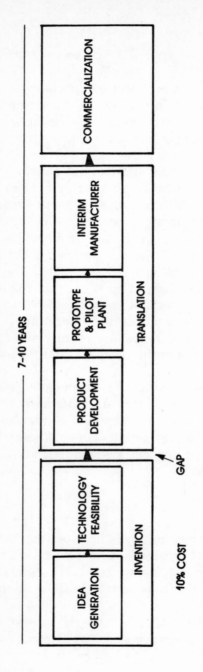

Source: Merrifield, U.S. Department of Congress

As the illustration demonstrates, the concept stage is only the first, and a very small step on the path to commercialization, and it must be supported by a significant expenditure of work and investment. Although the marketplace is eager to reward creativity, it also has learned to recognize the contributions of other participants in the commercialization process so that the technology can be launched successfully.

represents new knowledge, and that scientific knowledge truly has been advanced.

The extent of testing required varies widely, contingent primarily on the expertise of the researcher and the amount of money appropriated to this stage in the technology's life cycle. The research stage can mean just about anything, from primitive "garage" experiments to multi-billion dollar research efforts in stage-of-the-art facilities staffed by thousands of scientists.

Although the extent of research does not necessarily reflect the accuracy or validity of new knowledge, it does affect the perception of that knowledge in the eyes of the marketplace. If the research efforts completed at this stage in the life cycle are seen as shoddy or inadequate, the technology can be open to continual challenge by the scientific—or even the financial—community.

Sometimes, the new knowledge is rejected outright as untrue. Although some technology developers have the utmost confidence in their discoveries, peer challenge is important to monitor because the merest hint of question at the research stage can forestall the funding necessary to carry the technology forward to the next development stage.

When the research stage has been completed, the technology has advanced from concept form to a more tangible commodity. At this point, there exists some proof of the technology's viability: at a minimum, there is a written report which explains the original concept, the benefits derived in scientific terms, the processes and mechanisms used to demonstrate the scientific proof, and the results or measurements derived from the test.

What is important to know at this stage is that a simple "it works" does not constitute scientific proof. This is a major trap for entrepreneurs who are unfamiliar with marketplace demands for credibility: their testing methods have to be repeatable by any qualified researcher who has the proper equipment and materials. In fact, the ease with which testing can be repeated often requires its own unique technology so that others can verify the results of the entrepreneur's testing. As expenses mount for research verification, the technology required for testing can even exceed in value the technology itself. It is not uncommon for a technology breakthrough to be overshadowed by the value of the new industry created to verify the original technology.

To achieve maximum benefit from the research stage, experts say technology sellers should direct their testing toward commercial use because the validation process means more when it is applied to a practical solution. Although this can mean moving into the development stage prematurely, many scientific proofs fail the marketplace test because the extra field test step is omitted.

When this omission leaves only laboratory results on which to base the technology's progress, the results often can be considered meaningless by valuators. That is because scientific proofs at this point at least should attempt to establish economic proofs for the viability of eventual commercialization. When this all-important question is left unanswered at the research stage, it will have to surface later when even more development dollars have been spent. So the earlier non-commercialability is established, the less costly is the error for the developer.

It is also important to realize that although research may never yield results that actually prove the original hypothesis, many technologies reach the marketplace despite a lack of clarity about the scientific principles on which the technology is based.

Some technologies can even be commercialized without complete or even partial proof. But the inability to achieve certainty about the scientific basis of the technology usually leads to compounding, and expensive, problems in later stages of the life cycle.

So it definitely is preferable for the research stage to be completed successfully before moving on to the next stage, which is the development stage. If successful completion of the research stage does not occur in sequence, it can retard or even impede the commercialization and/or transfer of the technology altogether.

At the research stage, a technology can be transferred more easily; this is particularly true if the technological concept is viewed as the best solution to an acknowledged problem in the marketplace. Higher prices can be demanded in technology transfer at this stage if the buyer has an existing awareness of the market problem which the technology addresses, and he can envision how the technology solves that problem.

At this point, the developer can elect to transfer or to wait until after the development stage, when even more profit is attainable because the technology is that much closer to commercialization. He also can transfer all of the technology after the research stage, or he can commit to the responsibility of completing the economic proof stage before giving up all rights to the technology in exchange for a fee.

Development

Like the research stage, the development stage tests the hypothesis underlying the technology; then it takes proof one step further by testing the economic viability of the technology. In other words, it determines if the use of the technology in its final form will prove to be a net benefit and yield a profit to the entrepreneur-as-developer. These benefits are measured in terms of profit, competitive position and social welfare.

Because the acid test is actual market performance, it is impossible to conclude the economic proof of the technology successfully before the manufacturing, distribution, sales and use stages are concluded. But the costs associated with concluding these stages are usually steep enough that it is prudent to make predictions about the feasibility of making a further investment in the technology.

In order to prognosticate accurately, most technology developers create prototypes in the development stage. When the product is a technology, the prototype usually consists of computer simulations.

The use of models or prototypes in the development stage provides at least some feedback about what it will cost to commercialize the technology. So development-stage testing is more accurately a fact-gathering process in which more specific attempts are made to estimate the costs of each stage of the technology's life cycle in order to complete commercialization.

When the development stage is completed, the technology holder has an even more enhanced product, including a valid description of the technology, projected benefits from its commercial use, some degree of scientific validation and an economic estimate of the technology's market performance. In addition, the manufacturing costs have been reduced to a near-certainty and the probable performance of the technology in its final form are known.

Another aspect of the advanced testing characteristic of the development stage is the marketing of the commercialized technology: what price will the ultimate user pay for the opportunity to benefit from the technology? The classical marketing problem at this stage is determining who the user is, how many users there are potentially, what benefits can be expected from the technology, how much value is perceived by the user, and what price is fair to both the buyer and seller of the technology.

So marketing considerations at the development stage can be reduced to the technology's cost/benefit ratio, or making sure that total commercialization costs are less than the price asked of a potential user.

Development-stage testing seldom proves with any certainty that technology in the marketplace will yield a net benefit or profit. But at least when the technology is only a small step forward—an improvement over existing products and services— the technology holder is assured that a market for the technology exists.

In contrast when the technology is considered a breakthrough or giant leap forward in scientific knowledge, the existence, size and longevity of the market all are open to question. So when the technology is a breakthrough, it has to reach the marketplace before cost/benefits can be proven.

When the entrepreneur has reached the development stage, it is good to review the myriad ways commercialization can be achieved because these options usually have quite different results when cost/benefits or profitability are measured.

Some technologies have such enormous economic potential that they become profitable under almost any marketplace circumstances. Other viable technologies, however, are so weakened either inherently or because of marketplace conditions that they may never become profitable for the owner. Most technologies fall somewhere between these extremes on the economic continuum; this means their economic viability is contingent on good technology management.

How extensive does the economic proof have to be at the development stage? It is most common for an acceptable economic proof to be based on a single application of the difficulty in completing economic proofs for every conceivable application of the technology. If the technology holder relies on a single positive proof, however, the importance of that singular proof is weighted because of the additional resources committed to advancing a single application.

Single-application proofs, then, suggest several caveats. First, the basis for selection of the single application tested has to be valid. Generally, single-application testing is based on the same guess work most entrepreneurs depend on throughout the technology development process: strategies about which the developer has the most knowledge.

Although this usually decreases the risk of chasing a "dead" application, hunches can ignore the "best" applications for the largest market or the highest profit potential.

Another caveat for the technology developer who relies on hunches is the question of what to do about all the other potential applications that never get tested. Most owners decide that when they dominate the marketplace with the first application of their technology, they will have the resources to develop other uses.

But this sequential approach is not usually acceptable to technology valuators because it relies too heavily on the presumed success of the first application. In addition, it allocates no resources for the transfer of the other applications. The solution, in most cases, is to transfer one or more applications before committing to the development of additional applications using the entrepreneur's own resources.

Manufacturing

This manufacturing stage is the portion of the technology's life cycle in which the technology is replicated so that more than one user can apply the technology simultaneously. (The term "manufacturing" refers to actual end-products; the term "publishing" applies to books, software and data.)

The goal of the entrepreneur at this stage is to make enough of the "product" to satisfy initial market need, and to accomplish the task at a cost that assures profitability over the long term. The goal of the product manufacturer is to "hand-craft" the first production run when only a small amount is required to meet market need, gradually increasing quantity over a period of years.

It is not uncommon for the manufacturing stage to be the most capital intensive of all stages in the life cycle. This is particularly true when unique secondary technology is required to complete manufacturing. As a consequence, there is great incentive for a technology transfer to occur before the manufacturing stage to avoid the capital investment that manufacturing requires.

After manufacturing, the technology is viewed more as a finished product than as an intangible advance in science. The remaining stages of the life cycle, therefore, appear to carry about as much risk as any other finished product—rather than the risk of an unknown technological advance. As a result, there is a disincentive to transfer prior to the manufacturing stage when there is an element of mystery about the technology that can decrease its value. If the entrepreneur can work through the manufacturing stage, the marketplace will determine that he has borne all the significant risks and he can bargain in the transfer more effectively.

One alternative to capitalizing the manufacturing stage is to subcontract this stage, which is a form of transfer. The entrepreneur can enter into an agreement with an existing plant or job shop, for example. Typically, a fee is paid for manufacturing which covers manufacturing costs plus a fixed profit markup. This alternative allows the entrepreneur to gain several of the benefits of transfer without having to give up all the rewards. Once the manufacturing stage is completed, he can begin the distribution or sale of the product, or transfer away.

At the manufacturing stage, the owner is concerned only with the single application of the technology that he has developed, although this application may encompass several end-products. When this stage is completed successfully, the entrepreneur is expected to have a finished technology which is ready for sale to and use by an individual or corporate entity.

Distribution

The distribution stage, which covers the transportation of the technology product from the manufacturer to the retailer, can utilize several tiers of re-sellers; or distribution can skip the retailer and go directly to the end-user. When distribution applies directly to the end-user, it becomes partly a function of sales, which is described more fully below.

Transportation of the technology from Point A to Point B is a key issue because it can prove to be a significant obstacle, contingent on the characteristics of the product. The costliest obstacle for the entrepreneur arises when the cost of transportation exceeds the cost of manufacturing. Another transportation obstacle arises when governmental policies forbid or restrict shipment of the technology across international boundaries for national security or other reasons.

Despite these obstacles, the distribution stage usually is easy to complete successfully in comparison to other stages in the life cycle. Distribution does gain some importance when the technology is packaged with other products and/or services, which increases the potential value of the technology as a component in a larger transfer package.

Technology transfer becomes an important issue at the distribution stage when manufacturing is more profitable as a strategy to overcome the costs of transportation and import tariffs. When this stage has been completed, the technology is presumed to be in final form and ready for sale to the end-user. So in terms of risk, the distribution stage generally is not a problem for the technology holder if he has completed all the preceding stages in the life cycle. If the owner delays technology transfer until the distribution stage, he is rewarded with an even bigger slice of the profit pie for performing the distribution tasks and covering the carrying costs of inventory from manufacture to sale.

The distribution stage is a common point in the technology's life cycle for transfer: the technology can leverage off streams of existing commerce as other products are delivered to market. And the shared distribution costs can meet or exceed the desired cost objectives for an established distribution network.

A transfer at or before the distribution stage compels the owner to face the realistic salability of his technology end-product. This is particularly true when all life cycle stages are completed, because even with successful completion it is tempting to believe a transfer will be profitable.

The ultimate profitability of the technology transfer can not be known, however, until the sales stage, which is considered the litmus test for any technology. In fact, some technology sellers fail to acknowledge that the end-product is not commercial until it actually bombs in the marketplace.

That is why many owners—unable to judge accurately the marketability of their technology— prefer to transfer the technology when the responsibility for distribution and/or sale can be placed on a third party. If no one is willing to buy the technology at a wholesale price, however, then it most likely fails the commercial test unless the owner develops a successful strategy for direct marketing the technology to the end-user.

Sales

The sales stage is the final stage of the life cycle of the technology or product over which the owner has at least some control. At this point, the technology or product is considered complete, with all of its primary characteristics developed, proven and in place for use.

It is at this stage that the technology or product must be transferred from the entrepreneur to the end-user or consumer.

Because there are no more responsibilities for getting the product to market—it has already arrived—the term "technology transfer" is seldom used at this point. But it is important to realize that the implementation of technology transfer methods can support sales activities very effectively at this stage.

The objective of the sales stage is to complete a transfer of the technology or product if that has not already occurred. If all previous stages have been completed, sales activity consists primarily of educating the consumer about the benefits of the technology.

The major risk at this stage is potential buyer rejection or challenge. A secondary risk is lack of availability due to operations problems, or lack of sufficient sales volume to cover the development investment made in previous stages. So a transfer has to occur at this stage or the technology will be an economic failure. Although a transfer at the sales stage does not involve a transfer of responsibilities for commercialization, it is still perceived as a risk for the buyer.

Use

In the use stage, the technology is actually consumed— an end-user takes some action with the technology in order to derive a benefit. It is at this point that the technology is most highly valued in terms of improvement or solution to a problem experienced by the end-user.

At this point, many entrepreneurs believe they have no additional tasks to perform because the technology has been transferred/purchased. But this is usually not true because more often than not the end-user does not have sufficient knowledge about the technology to maximize the benefits that are possible through the use of the technology.

Sometimes the end-user does not even understand how to use the technology. So it is considered incumbent upon the technology holder to provide a solution to the usage or maximization potential. Most entrepreneurs develop instructions, user manuals, training programs and other forms of instruction to answer user needs.

This education and use support frequently requires additional financing in order to expand on the knowledge required to develop the technology originally and then implement it. The primary risk at the sales stage is that the end-user decides he has not received all the benefits possible from the technology. This risk can lead to an outright rejection of the technology, demands for refund, failure to make repeat purchases or failure of word-of-mouth marketing.

Straight Linear Progression

The development and commercialization of a technology seldom proceeds from the concept stage to the use stage in a straight- forward, linear fashion. In the real world, there can be fewer stages or more; some stages are repeated; and some stages are the ground for a total shift in developmental direction.

Technology development as portrayed in the description above illustrates the development of a single item of knowledge. But it is not uncommon to develop multiple technologies within a single life cycle for one application of the technology. In fact, the more dynamic the original technological concept is, the more likely is a multiple application development process that overlaps several industries for sale in multiple markets.

The only unique requirement in this case is that each separate application of the technology requires a separate economic proof. The resource requirements in terms of financing, facilities, expertise and equipment can vary radically from stage to stage, representing completely different investment decisions. The stages are presented as discrete, each distinct from the other. However, real-world development tends to meld the distinctions among stages, some of which are never completed.

Go/No Go

Some, if not most, technologies never make it through an entire developmental life cycle; they tend to falter at the two most important barriers: the scientific proof and the economic proof. Unfortunately, not all entrepreneurs are wise enough to halt development when these proofs fail. Some owners ignore their own test results and proceed with the next stage of development, wasting valuable resources pursuing a technology that has little or no market value.

So effective technology management requires constant evaluation of the technology in order to justify continuing investment in time, money and effort to get the technology to market. Technology management is even more crucial when the technology can not be validated with scientific or economic certainty. A "go/no go" analysis always should follow scientific and economic testing so that progress can be matched realistically with the developer's resources. (See Chapter 5 for a full description of technology management.)

Marketing

Marketing is not usually considered to be a life cycle stage because it is more accurately a continual assessment of user needs, of the technology's capability to satisfy those needs and of the further completion of the economic proof. This requires the marketing process to be initiated during the concept stage and to play a vital role throughout every succeeding stage. (See Chapter 6 for a full description of technology marketing.)

Calendar and Time Lines

It is difficult to create a calendar or time line that applies to the completion of all stages in the technology life cycle (See Chapter 3 for a rough calendar outline of timelines for each stage in the transfer process.) Some technologies can be developed so rapidly that the seven stages are completed almost simultaneously with no obstacles.

Many more technologies, however, struggle at plateaus in every stage before moving forward to the next stage. It should be emphasized that a significant number of technologies fail to complete all seven cycles, and that the omission of any one step—particularly the proofing stage—substantially increases the chance of failure to commercialize a technology.

Points of Transfer

Theoretically, technology can be transferred at any point during the development life cycle, that is before, during or after any one stage. In terms of profitability and ease of transfer, the more complete a technology is—the closer it is to the marketplace—the easier it is to complete a transfer because fewer risks and costs remain for the buyer to carry.

There are several guidelines that can be used to decide when in the technology's life cycle to transfer relative to potential profit, risk and resource requirements (see Chapter 7 for a fuller description of these formulae):

1. In most industries, the earlier in the life cycle technology is transferred, the less money the seller will receive. Some buyers attribute absolutely no monetary value to the concept stage even if it is the "heart" of the technology.

2. Typically, profits are distributed among the life cycle tasks—as points of transfer—based on the level of risk and capital investment required to perform that task.

3
The Technology Transfer Process

Approximately 90 percent of classical transfers in both the private and public sectors occur between the R&D stage and the manufacturing stage of the life cycle of the technology.

The underlying premise of technology transfer is to shift the responsibility for commercialization of a technology from one party or entity to another. This transfer represents not so much an event as it is a protracted process that varies by industry, company and technology. Sometimes, the transfer process itself is so individualized it varies almost by entrepreneur. This process can have a myriad of forms, but each transfer has certain commonalities that are illustrated best by the following tasks that must be performed by the entrepreneur as technology seller:

- The identification of technology: inventory, analysis, control
- The identification of prospective markets: industries and buyers
- The validation of technology: testing and packaging
- The sale of technology: presentation, negotiation, deal
- The actual transfer of the technology: structure, price, risks, ownership

Every transfer also includes buyer activities that share certain commonalities:

- The identification of the technology
- The identification of the inventor/rights holder
- The validation of the technology: due diligence
- The purchase of the technology
- The actual transfer of the technology

At first glance, these activities seem to be closely related by their nature—and they are. However, the difference in perspective between the buyer and seller dramatically affects the nature of the transfer activity, the difficulty both parties can expect to encounter in the transaction and the legal importance of each step.

Because these transfer steps are time-consuming when performed diligently, transfers are transacted by sophisticated parties who intend to perform each step only once. But in some cases, prudent business practice suggests that one or more of the steps be repeated in order to address more thoroughly the risks assumed by each party as a result of incomplete scientific and economic proofs.

Seller Identification of Technology: STEP ONE

From the entrepreneur's point of view as seller, his technology is a solution to a prospective buyer's problem; so the identification of the technology in the section below is approached from the seller's perspective.

Inventory

The first step in the identification of the technology is to ascertain exactly what the inventor has to sell. This is sometimes referred to as "taking an inventory," or identifying the exact nature of the technology, analyzing its characteristics and attributing value.

Although taking inventory seems like an unnecessarily obvious activity, failing to identify a technology by taking inventory is not that uncommon an omission by entrepreneurs for a number of reasons. One is that the creation of the technology may not have been planned; it may have evolved as a by-product of another research objective. Or

Chapter Three

The Product Life Cycle and Its Relationship to the Innovation Process

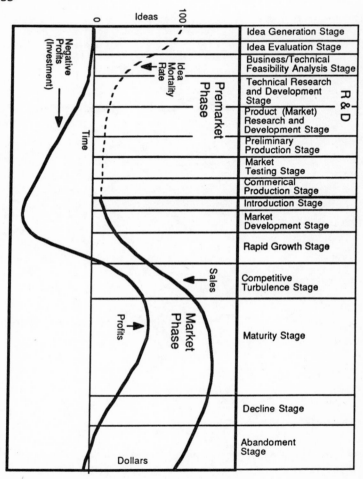

Source: The Innovation Assessment Center

Technology entrepreneurs can create a more detailed commercialization plan by studying how the technology innovation process is linked to the product life cycle and the investment cycle before attempting to validate the technology concept.

Science, Technology, and the Utilization of Their Products, Showing Communication Paths Among the Three Streams

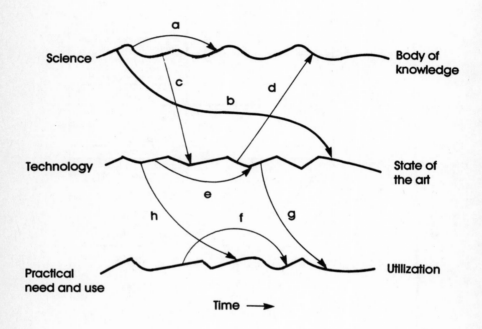

Source: Factors in the Transfer of Technology, edited by William H. Gruber and Donald G. Marquis, The M.I.T. Press

The first step in the development of a systems analysis for technology transfer is to review the relationship among the major transfer activities. This model indicates potential points of transfer and potential sources of conceptual generation. Here, transfer can occur among any of the three flows: from technology to use, or from science to technology, or within any one of the channels.

this specific application of the technology may have arisen from other applications or other technologies.

In the same way, the benefit derived from a technology may not be apparent or realized until a subsequent research event occurs. Also, a technology may have been shelved due to past unfavorable conditions.

So, however the technology came into existence, it is vitally necessary to identify it appropriately because the identification of the technology to be sold is framed in terms of what the seller has that is salable, and the subsequent value to both buyer and seller results from that identification.

For example, a software company publishes a package which measures seismic readings for the purpose of locating petroleum deposits. During the course of development, it realizes that a minor error in the software can distort the results significantly. Therefore, it creates an ancillary program which acts as a quality control mechanism by testing the validity of the report against the test data. This ancillary program has its own independent value and represents a completely separate product. However, unless the software company identifies the program and markets it independently, it is highly unlikely that any prospective buyer will ever know that it exists.

Analysis

After first-step identification of a technology is established, the seller must attempt to convey a solid understanding of how the technology functions by establishing its validity. The most effective criteria for establishing validity are the scientific and economic proofs described in Chapter 2. If the technology cannot be proven at the time of sale, the seller usually determines whether or not the proof can be completed, and if it can be, by what methods.

Analysis also determines what characteristics of the technology can be used to create legal protection for the new process, usually in the form of a patent, copyright or trade secret.

Finally, the analysis stage breaks down the technology into all possible common denominators as a listing of how many different ways the technology can be sold. This determination is framed in terms of the achievement of respective buyer and seller objectives. For example, in the case of a chemical compound that can absorb water rapidly and release it over time, the application of the compound appears to be infinite. It can be used to reduce irrigation in agricultural industries, to preserve food products in the food storage industry or to act as a dessicant in the plastics manufacturing industry.

Creation of Legal Control

The first step is concluded when the seller establishes legal control of the technology. This is vitally important because technology transfer cannot be accomplished without the creation of legal control at this stage. Rights of ownership serve as the only barrier to the free flow of information about the technology to the general public. See Chapter 8 for a full description of the legal aspects of technology transfer.

Buyer Identification of Technology: STEP ONE

A technology promises to be a potential solution to a problem from the buyer perspective, too. So the first step a buyer takes is to identify the problem he wants to solve before trying to identify the technology that will be the solution. To most buyers, in fact, understanding the problem concretely is the flip side of understanding technology.

With a clear picture of the problem in view, the buyer begins to hunt for ways to solve it and this is one of many credibility gaps for the seller: the buyer does not necessarily perceive a technology or any technology as an answer to his problem. Generally he peruses a great number of potential solutions, all of which may take a different approach to the problem, and are based on widely variant products, services or technologies.

Not only does this pose a problem for the seller, it usually confuses the buyer for a period of time before a technology sale takes place. That is because the buyer typically uses a broad approach to finding a solution, among such sources as databases, trade shows or exhibits, among others. He often encounters great difficulty at this stage because the relationship between the seller's technology and the buyer's problem is not readily apparent at first.

If the seller has identified, analyzed and created his technology proficiently enough to be recognized as a potential solution by the

Innovation Process

```
[Invention] — Concept Transfer — [Development] — Technology Transfer — [Commercialization]
```

Time frame — — — — — — — — — 7 - 10 years — — — — — — — — —

Cost — — — — — 10% — — — — — — — — — 90% — — — — — — —

Major transfer barriers

Source: Critical Success Factors for Commercializing Technology, F. Timothy Janis, Indianapolis Center for Advanced Research, Inc.

There are three major phases in an overview of the innovation cycle: the invention phase, the development phase and the commercial phase. As noted above, costs are heavily weighted toward the development and commercial phases. From the transfer point of view, two major barriers occur at the transition points: at the transfer of the technology from idea to development, and at the transfer of the technology from development to the marketplace.

Pace of Some Innovations

Technology	*Conception*	*Entry*	*Period*
Heart Pacemaker	1928	1960	32
Magnetic Ferrites	1933	1955	22
Transistors	1940	1950	10
Oral Contraceptives	1951	1960	9
Videotape Recorder	1950	1956	6

buyer, the buyer begins to examine the technology in terms of potential solutions to his problem, but not in terms of potential performance at this stage.

Say a company—as a potential buyer—wants a computer-aided design (CAD) system to have the capability to prepare architectural drawings. The CEO can begin the search for a CAD system by going to a known vendor of CAD systems or to a publication about CAD systems. But at this point, he has no specific knowledge about which vendor designs a system that most clearly meets his need, or about which vendor produces successful systems in the marketplace.

Seller Identification of Buyer Market/Industry: STEP TWO—PART I

In the second step, the seller identifies prospective buyers of the technology—individuals or corporate entities who ultimately can benefit from the commercialization of the technology to end-users in the marketplace. It is important here to distinguish between the direct buyer of the technology and the end-users of the technology to whom the buyer markets.

Buyer identification advances the transfer process in three ways: (1) it is the first step toward an economic proof because it shows value in terms of user benefits to a specific entity; (2) it helps pinpoint different applications of the technology, also in terms of end-users; and (3) it is the cornerstone of targeting key industries for current and future sales, and of marketing to existing and new end-users.

Because the process of technology transfer shifts the responsibility for commercialization to the buyer, it is crucial that the prospective buyer be capable of completing the tasks remaining in the technology's life cycle. Usually the best buyer comes from an industry that already services the prospective end-user.

It strongly indicates that the seller should begin the search for a buyer from within the buyer's industry. Once this industry is identified, prospective buyers can be found readily by name in that industry's trade and professional publications, and within its trade associations.

Chapter Three 49

Model For Linking Manufacturers with Technology Resources

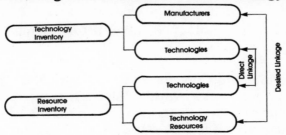

Source: *Technology Transfer and Regional Competitiveness*, Dr. Edward H. McMahon, University of Tennessee at Chattanooga

This model for linking manufacturers with technology resources mirrors a popular corporate model used to link a company's technology resources with its business plan; therefore, it is a useful picture for both buyers and sellers to review before setting technology transfer objectives.

Buyer Identification of the Inventor and Seller:
STEP TWO—PART II

From the buyer's viewpoint, the possibility of locating a seller with the right technological solution can seem as futile as finding a needle in a haystack. It may look doubtful that a technology even exists to fulfill the needs of the corporation. One approach used by buyers who want to locate sellers efficiently is to hire consultants who have much broader knowledge of and interaction with their industry. And there is a variation of this strategy when a technology cannot be identified after a long and fruitless search for the solution: the consultant simply bids on the third-party creation of an appropriate technology that will meet the buyer's needs. Another variation is to network with industry organizations that maintain valuable source libraries and use backup advisors.

Yet another approach is to contact experts within the relevant field of science, for example, experts from federal laboratories, university facilities or the private sector who can be identified in trade and technical publications. These experts sometimes also are technology transfer facilitators who serve as information sources, technology dealers and deal-brokers. (See Appendix A for source names.)

Seller Validation of Technology: STEP THREE—PART I

Scientific Proof

The cornerstone of technology transfer is the scientific proof that answers the question, "will it work?" Although many technologies reach the marketplace without answering this question adequately—the number of hair growth formulas and weight loss pills are a good example—they will not prosper and become profitable without scientific proof.

Testing is the most common form of proof, to find uses of the technology by which specific benefits can be identified and measured. The proof must involve tests that are repeatable with consistent results by third parties. Entrepreneurs should take note that the primary reason testing fails is because it is conducted in a laboratory under conditions that do not exist in the marketplace.

The conditions under which testing occurs must be realistic and valid in order to validate the proof; validity is determined when the technology is in actual use.

So in order to achieve maximum credibility with regulatory agencies, investors, buyers and the marketplace, testing should be performed by a third party who is not controlled by the buyer or seller, and who has a reputation for scientific integrity.

The caveat, however, is that the use of third-party researchers may be unrealistic because no one else is qualified to test the technology; or disclosure to a third party may significantly jeopardize the competitive placement of the product in the marketplace.

In any case, the seller should anticipate and satisfy the desire of the buyer to review test results and to recreate the test results for definitive proof of validity.

Economic Proof

The other component of validity answers the question "what is the technology worth?" All too often this question is relegated to a consideration of maximum possible gross revenue from the sale of the technology to any or all possible end-users. This is not an improper form of validity valuation, but it does fall short of providing enough useful information to complete the proofing process.

The best formula for pricing a technology is to estimate conservatively the revenues that can be generated from successful commercialization of the technology, less the costs of getting the technology to the use stage. This formula is:

$$\text{Net Value} = \text{Gross Revenue} - \text{Gross Expenses}$$

Gross revenue represents the cumulative retail revenue generated from the sale to a user. Gross expenses are the cumulative costs of taking the technology from its current stage to a final form, including all the costs of distribution and sale as well.

There are a multitude of other workable formulae that measure the value of a technology, and the better formulae also take into consideration the risks borne by the entrepreneur in commercializing the technology. That means a technology which requires greater resources of time, money and labor is considered inferior in value to a technology that has the same net value with half the investment.

Buyer Validation of Technology: STEP THREE—PART II

The buyer's process for validating the technology—analogous to the seller proofs—is the buyer's "due diligence" stage. The buyer now begins to qualify the technology with regard to its ability to resolve his target problem, as a form of scientific proof. However, to the extent that seller testing or proof has sufficient credibility, additional buyer testing may become unnecessary.

Examples of third-party proofing sources:

- Private Sector:
  SRI International, 333 Ravenswood Avenue, Menlo Park, California 94025-3493, (415) 859-4771
- Academic Sector:
  Colorado State University, P.O. Box 483, Fort Collins, Colorado 80522, (303) 482-2916

- Public Sector:
  National Institute of Standards and Technology, Laboratories and Centers, Route 270, Gaithersburg, Maryland 20899, (301) 921-1000

If the technology appears to function the way it was intended to, the buyer then must determine the price at which acquisition is justified—unless the seller has proofed this issue well enough to eliminate the need for the buyer to seek independent verification. The valuation formula is:

Average Benefit to User × Projected Number of Users − All Commercialization Costs = Projected Market Price

For example, if a solvent is developed to remove ink stains from men's dress shirts, a male consumer could benefit from the product by avoiding the cost ($25) of purchasing a new shirt. A reasonable projection of prospective users is 10 million male shirt-wearers. Total projected benefit from the use of the solvent is, therefore, $250,000,000. This amount is the maximum potential revenue from the sale of the solvent. From this amount must be subtracted all of the costs of commercializing the solvent. If the cost per bottle is $10, then the total cost of commercialization is $100,000,000. This creates a net value, net profits or projected maximum price for the technology at $150,000,000.

But this example illustrates how many entrepreneurs use a pricing formula: they simply multiply the numbers without using a heavy dose of realism. So the entrepreneur must always consider the many additional factors that impact the pricing formula. One common factor is the presence of a competing technology. Even if the costs are the same for two products, the sale price of $25 would be reduced in a price war, for example, between two competing products. If the competing technology is available to consumers at a lower price, the solvent may have no value at all. Another factor is that the user rarely pays the full price for the benefit gained by using the product. In this example, the user has the option of spending $25 for the new solvent or just buying a new shirt. Reality favors the new shirt, which would force a reduction in the sales price to make purchasing the solvent the more desirable action.

The Sale of the Technology: STEP FOUR—PART I

Presentation

The successful—that is, profitable—sale of a technology depends to a great extent on the professionalism of the seller's presentation, for which all of the marketing rules must apply in the face of what the entrepreneur safely should presume is an indifferent marketplace.

As described above, the buyer of a technology typically reviews all possible solutions to his problem, which creates competition for the seller and the technology unless the technology is so unusual that there is no other solution for the buyer's problem.

So it is safer for the seller to presume that not only is the buyer not interested in this technology, but that he is not even aware of its existence, or that the need for a solution is unperceived. The seller's task is, first, to create an awareness of this technology and its potential, and then to promote its sale in answer to a specific set of buyer parameters.

Even if the buyer is aware of the need to solve a corporate or marketplace problem, his skepticism creates a "show me" attitude to the seller. At this point, the seller must be prepared to show off the technology in its best light with both scientific and economic proofs. The buyer must be convinced that the technology works and that its commercialization will yield significant results for both parties.

Most sellers tailor the presentation to each prospective buyer because a generic presentation of the technology seldom generates much interest. This means the economic proof also must be framed in terms of the buyer's specific application of the technology; and it takes into consideration all other alternatives available to the buyer, including a no-action scenario.

One aspect failing in many seller presentations is the amount of detail included: it is never enough. Although the buyer may appear to be unwilling to investigate the technology on his own, it should be noted that the buyer may be looking at several technology options, and does not have the time or money to take an indepth look at each one. But he will want the data, nevertheless. Therefore, the responsibility is on the seller to satisfy all buyer information needs, including data on how the technology fits the buyer's business plan, the competitive strategy and the product line.

Seller Checklist: Anticipating Buyer Needs and Questions

- Does it work? (requires documentation)
- How well does it work? (requires prototype, test marketing and documentation)
- Is there clear ownership? (requires proof of inventorship or patent application, and existence of intellectual property right)
- What is a fair price? (document with formula)
- What is the market demand? (requires research documentation)
- What are competitors doing? (requires analysis of the availability of alternative technologies)

Negotiation

Standard negotiating techniques often are portrayed as a power game in which winning is based on how tight the terms of the final agreement are constructed. This may work in some contract negotiations, but these tactics are not recommended in technology transfer negotiations. In fact, several aspects of a technology transfer demand great flexibility in the way terms are set in order to avoid a complete collapse of the deal structure.

Technology transfer negotiations generally need to be framed in such a way that the terms accomplish the objectives of all parties, not just the buyer's terms or the seller's terms. Even when one party has a clear advantage in bargaining power over the other party, caution should be exercised in the use of that power, and the reason is simple.

Technology transfer is a process that is complex and difficult to execute successfully. Many times the deal fails even when all participants in the transfer are in agreement about the terms and commit themselves to its completion.

This failure can result from a number of factors, all of which exist beyond the control of the participants, including the emergence of a competitive technology; a drastic change in market demand; participation in a trade mission in a politically volatile market; the death or divorce of a participant in the potential transfer; or a change in regulation or legislation. So it is prudent to believe and act on the premise that this risk of failure actually is increased during the stress of negotiation.

Instead of power bargaining, the principal purpose of negotiation is to identify and solidify the objectives of each party. This allows everyone to estimate more realistically whether or not the deal in its entirety can be completed. What generally happens is that remaining risks—those not eliminated in the negotiations—are relegated to one or more participants who assume responsibility.

Deal

In order to sell a technology, the transfer documentation must contain several common variables that are fully described, including:

- The deal structure/mechanism
- Ownership of the technology
- Final price
- Method and amount of payment
- Assumption of risks
- Time lines

Structure/Mechanism

The deal structure selected by the buyer and seller actually determines how the technology will be transferred. (Transfer mechanisms are more fully described in Chapter 4.) Typically, the seller selects the structure that best accomplishes his objectives as determined in the pre-planning stage. (Planning components are described in Chapter 5.)

Because ownership means control, the seller transfers some or all of his legal rights to the technology to the buyer in a transfer. (Types of rights and the manner in which they are created are outlined in Chapter 8.) But instead of focusing on rights, price often gets undue attention in a transfer negotiation. That is because to the seller, the transfer of responsibility to the buyer is a chance to cash in his hard-earned chips, walk away from the labor-intensity of developing the technology, and pay off the long-term investment he has made.

Therefore, the seller usually tries to get top price for the technology without recognizing buyer risks and the time yet to be invested by him. To ensure a transfer, the seller price should represent a reasonable profit on tasks already performed by the seller and include a premium for his creative input to its development. The pricing formula is repeated here:

Average Benefit Per User × Number of Users − All Commercialization Costs = Potential Market Price

Then the market price has to be allocated among all commercialization tasks in order to price the technology at any stage of the life cycle at which the developer chooses to sell it.

Referring back to the ink solvent example, the maximum price for the technology depends on the point at which the solvent is sold. The earlier in the life cycle and the less the entrepreneur has invested, the lower the price the entrepreneur will receive in the transfer. As illustrated below using a percent of cost method, the price for the technology after completing the research phase is $1,285,000, which recaptures $510,000 in costs and allocates a profit of $765,000.

Allocation of profits based on a percent of cost does not reflect fairly the risks incurred in each stage of the life cycle. Clearly, some stages carry greater risks than others. The research and development stages are always high because the scientific and economic proofs have not been completed.

Therefore, pricing technology also should carry a risk factor to allow a fairer allocation of profits and justly reward entrepreneurial effort. As the second illustration demonstrates, the recognition of risk profoundly impacts the profit allocation.

Pricing Technology At Different Stages By Allocation of Price According to Percentage of Total Commercialization Costs

Pricing Technology at Different Stages by Allocation of Price According to Percentage of Total Costs of Commercialization

Stage of Life Cycle	Costs of Commercialization	Percentage of Costs	Allocation of Profits	Asking Price	ROI
Concept	10,000	.01%	15,000	25,000	50%
Research	500,000	.50%	750,000	1,250,000	50%
Development	500,000	.50%	750,000	1,250,000	50%
Manufacture	46,000,000	46.00%	69,000,000	115,000,000	50%
Distribution	15,000,000	15.00%	22,500,000	37,500,000	50%
Sale	37,990,000	37.99%	56,985,000	94,975,000	50%
	$100,000,000	100.00%	$150,000,000	$250,000,000	50%

Pricing Technology at Different Stages by Allocation of Price According to Percentage of Total Costs of Commercialization and Percentage of Relative Risks

Stage of Life Cycle	Costs of Commercialization	Percentage of Risk	Percentage of Costs	Allocation of Profits	Asking Price	ROI
Concept	10,000	10.00%	.01%	7,507,500	7,517,500	75075%
Research	500,000	30.00%	.50%	22,875,000	23,375,000	4575%
Development	500,000	20.00%	.50%	15,375,000	15,875,000	3075%
Manufacture	46,000,000	10.00%	46.00%	42,000,000	98,000,000	91%
Distribution	15,000,000	10.00%	15.00%	18,750,000	33,750,000	125%
Sale	37,990,000	20.00%	37.99%	35,992,500	73,982,500	95%
	$100,000,000	100.00%	100.00%	$150,000,000	$250,000,000	50%

At this point, price and payment should be negotiated separately. The seller usually prefers to take payment in a lump sum, unless tax issues make it prudent to spread out payments over time in order to avoid paying maximum taxes in a higher bracket by taking the payment in a lump sum. (Accounting and tax issues are described in Chapter 7.)

The key issue here is that full payment by the buyer upfront protects the seller from the risk of the buyer's subsequent failure to honor the payment terms of the contract.

There are other risks inherent in the deal at the time of the transfer. The seller may want to force the buyer to take the technology "as is," including any remaining risks such as an inability to commercialize the technology despite further development efforts, disappearing market demand, changes in the regulatory environment or the availability of superior technology at a lower price. This is the lump-sum-payment premise which precludes future payments based on future success in the marketplace.

The last primary risk is pinpointing when the transfer will occur. A timeline can be quite definite in a straight transfer; but it is nearly impossible to determine a calendar sequence in most transfers. When it is possible, time lines should be established even if they serve no other purpose than to outline the possibility of a deadline.

The seller should guarantee that the scientific and economic proofs are completed and that the due diligence needs of the buyer have been anticipated; only then is it appropriate to present the technology package and solicit the transfer.

The caveat in the timeline below presumes the existence of unlimited funding and the availability of all necessary human resources. Without these resources, the timeframes can triple or even quadruple before commercialization occurs.

Projected Transfer Timeline:

- Scientific Proof: If only a simulation or forecast, it takes on average from one to six months. If a prototype is created, it can take from one month to two years for prototype manufacture.

- Economic Proof: Can take from one to six months for market testing. Proof based on actual sales in a limited test market adds another six months to one year to the calendar. A true economic proof occurs only after the technology goes to market.
- Market Study: Identification of each end-user or group of end-users who may benefit from the use of the technology and a projection of benefit for each end-user. This can take from three months to one year.
- Industry Study: Identification of industries that service the defined end-users and the members within those industries. This step takes approximately one to three months.
- Preparation of the Presentation Package: This step includes scripting, role modeling, and development of printed and audio/visual marketing materials in preparation for presenting the technology to potential buyers. This step ideally is repeated on a tailor-made basis for each potential buyer. It can take one to three months.
- Presentation Per Buyer: Contacting the prospective buyer, setting up an appointment, introducing the technology in presentation and following up can take from one month to two years, contingent on access to the buyer, the buyer's "seasonal mentality" and the buyer's location, especially if overseas.
- Negotiations: A discussion of deal terms takes one to six months in a typical transfer.
- Transfer: Actual exchange of information, documents, prototypes, personnel and training services relevant to the technology can take from one to six months.

Commercialization Concept

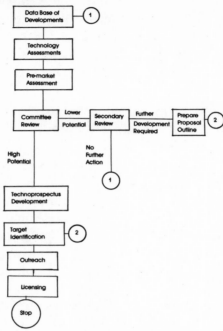

Source: *Critical Success Factors For Commercializing Technology*, F. Timothy Janis, Indianapolis Center for Advanced Research, Inc., Indianapolis.

Entrepreneurs should view technology transfer as a commercialization concept and a business opportunity that is bottom-line oriented rather than as an opportunity to develop a novel technology further.

The Buyer's Deal: STEP FOUR—PART II

In a typical technology transfer, the buyer negotiates the same terms as the seller in selecting a transfer mechanism, establishing a price, setting timelines and allocating transfer responsibilities; but the buyer's deal highlights different aspects of each term based on his corporate objectives.

First, it is most likely that the buyer will agree with the seller on the selection of the deal mechanism for the transfer, or there is no deal. As described later in more detail, a variety of mechanisms exist, but rarely will there be more than a few viable alternatives; typically, one structure is superior because it clearly accomplishes the objectives of both the buyer and the seller.

The buyer is more concerned with ownership issues than is the seller. That is because a substantial investment usually remains in the development of the technology and the buyer cannot risk a lost investment in a technology that belongs to someone else or which cannot be controlled through ownership.

Price is also a special buyer concern for the same reason: if a price based on the presumption of success in the marketplace appears too high, the buyer usually tries to establish a lower price that he can afford to lose.

Usually the buyer tries to arrange a minimum payment at the time of transfer with additional payments tied to subsequent market performance. This serves two purposes: it reduces risk if the product fails and it defers payment until revenue can be generated by which to makes the payments.

A typical minimum-payment formula suggests that the minimum payment made by the buyer at the time of the transfer should cover at least the cost of the transfer, and it can go as high as the total seller investment in the technology to date. Subsequent payments to the seller are a portion of profits or revenues as negotiated between the buyer and seller.

The buyer also tries to pin some of the risks on the seller. His reasoning is that if the technology proves unworkable or unprofitable, the seller should not walk away with a profit at the buyer's expense.

These risks can include claims for product liability injuries, the buyer's investment in the technology (the buyer will recover all of his costs first, before payment to the seller) and/or any infringement of someone else's technology (if the buyer and seller unknowingly infringe on another technology, any resulting lawsuit by the competitor will be paid by the seller; in this situation, the infringee forces the seller to buy a license to his technology).

The buyer also seeks definite deadlines for actual transfer and for when the seller must discharge his obligation to remit documentation and training to the buyer; but he is much less concerned about when his subsequent payments are due the seller, when his benchmark performance deadlines to commercialize the technology occur, and/or when he will advertise the technology, or enter new or foreign markets, for example.

Seller Transfer Technology: STEP FIVE—PART I

The transfer of a technology is perceived more appropriately as the transfer of knowledge, and this is never more evident than at the time of transfer. Transfer is never as simple as the delivery of a tangible product that can be picked up and used at will by the buyer.

Even the most common technologies require instruction for installation, use and maintenance. Early stage technologies usually require even more indepth assistance and effort in order to complete the transfer.

Typically, follow-on instruction and training are contracted in the form of technical consultation to transfer the data from the seller's head to a self-evident form of information like a training manual. The inventor must continue to explain his concept as use is developed by the buyer.

Contract terms include seller availability to the buyer, participation in subsequent technology review, decision-making input and training of buyer personnel.

The transfer itself is determined to a significant extent by the structure or mechanism selected by the buyer and seller. After that, the primary aspects of the transfer are keyed to the elements of knowledge that comprise the technology.

These also are keyed to aspects or characteristics of the technology that benefit end-users. If a cooperative R&D venture is planned as a transfer mechanism, the nature of the buyer/seller relationship arises from the specific characteristics of an R&D model.

For example in a cooperative R&D venture, the buyer will want a mechanism that provides even more access to seller input and knowledge after the transfer than in a final sale or licence to compensate for the lack of shared participation in the future tasks of commercializing the technology that exists in a sale or license mechanism.

The transfer of the technology is not considered successful unless steps are taken to provide the "secondary knowledge" about how to build and use the product. There is a caution at this point, however. Many aspects of a transfer are no more than terms of education and instruction that include what the seller knows about the technology and what he can do to take the technology to market. Often, the seller has a tendency to treat this data as too basic to have to explain in enough detail.

Unfortunately, this attitude says to the buyer, "here is the technology, what more do you need?" and it is an attitude that can prevent the transfer from taking place. The buyer not only wants all pertinent data, he wants, as a rule, all the support and follow-on consultation he can get. (See Appendix C for specific terms in an actual technology transfer contract.)

Buyer Transfer Technology: STEP FIVE—PART II

On his end, the buyer relies heavily on the deal mechanism selected as the tool that will solve his corporate problem; sometimes that means the buyer wants to take a dominant, "I'll do it all" attitude.

One reason is that the technology may appear to be vague and undefined, causing him to take extra precautions to assure that all available knowledge about the technology is transferred. He will want to review any documents that have been prepared by the seller that relate to the technology, including positive and negative laboratory reports, correspondence with other technology developers, professional papers and articles for trade publications.

And the buyer often puts greater emphasis on secondary knowledge because he wants to know more about how to make the technology than how it actually works. So instruction and support are critically important for a smooth transfer. In fact to get the seller's full attention until the transfer is completed, he often contracts for the services of the seller in order to gain as much knowledge and expertise as possible.

4
Technology Transfer Mechanisms

About 80 percent of all classical private and public sector transfers are licence transfers, which are considered the most efficient, straightline method for transferring ownership.

Rarely does technology transfer resemble a straight sale or an instantaneous event in which responsibility for the commercialization of the technology immediately shifts from one party to another. Most

Functional Transfer Model II

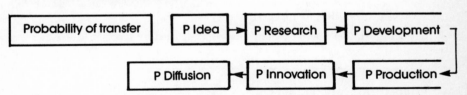

Source: *Factors in the Transfer of Technology,* edited by William H. Gruber and Donald G. Marquis, The M.I.T. Press.

This diagram indicates the sequence of events within the company structure of a potential recipient or user of existing technology. The entry point for a new idea is the probability that someone in the firm has a new piece of technical information to examine which will lead to commercialization.

transfers are very complex, involving many individuals or entities over a long period of time.

Several workable mechanisms have evolved over time by which technology is transferred most effectively. These vehicles are described below relative to:

- Deal Structure
- Implementation of the Transfer
- Seller and Buyer Roles
- Advantages and Disadvantages

Technology Transfer Mechanisms

Model	Principal Activity	Example
Agent	One-on-one assistance	Agricultural Extension Service
Linker	Linking developer with user	Federal Laboratory Office of Research & Technology Assessment
Disseminator	Information retrieval and distribution	NASA Industrial Applications Center
Licensor	Legal transfer/acquisition of intellectual property	Center for Utilization of Federal Technologies

Source: *Critical Success Factors For Commercializing Technology*, F. Timothy Janis, Indianapolis Center for Advanced Research, Inc., Indianapolis

These categories of model transfer participants are designated by function in the transfer process.

Less Than Complete Transfers

Sometimes a transfer is a temporary structure by which the owner of the technology shifts responsibility for only a single stage of the life cycle or for one task. This is referred to as "sub-contracting." A sale or license has been transacted in which the buyer receives certain rights; but the transfer is not complete because it represents a delegation of tasks.

Chapter Four

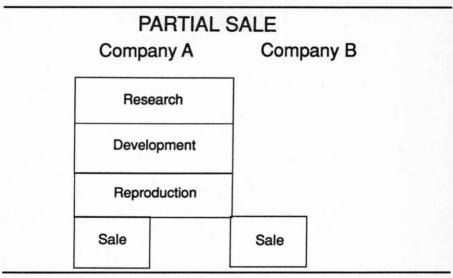

Other transfers are partial because the owner of the technology retains shared responsibility for the performance of a life cycle stage or a task. Both of these cooperative arrangements also involve the transfer of rights. Any form of transfer EXCEPT the straight sale or assignment is an incomplete form of transfer, including the license transfer, publication transfer or cooperative R&D transfer.

Straight Sale

This is the simplest form of technology transfer, if it can be said logically to exist at all. As the name implies, this transfer represents a complete shift in responsibility; the owner of the technology divests all responsibility for its commercialization to the buyer.

The seller also transfers all legal rights held in the technology. If these rights have not been perfected through registration and filing, then the seller must transfer rights in the technology as a trade secret.

This transfer can occur at any time in the life cycle of the technology; however, it is likely to occur at the end of the cycle when many of the unknowns have been answered and the technology becomes a "black box," or a more tangible commodity.

In a straight sale, the seller must bring to the trading table all

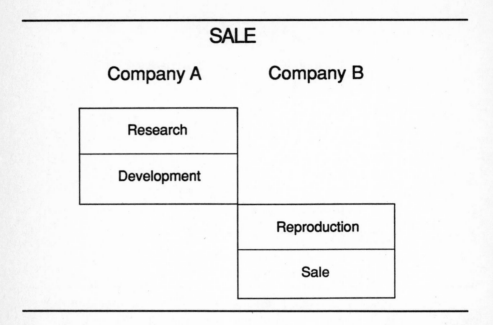

relevant information regarding the technology; no data necessary for the buyer to complete commercialization of the technology can be retained by the seller.

The buyer, too, must be sure to obtain everything that should be sold, including all documents about the technology, engineering diaries, test results, market surveys, cost projections and business plans.

The buyer is, in essence, acquiring an opportunity based on the knowledge of the seller at the time of transfer. The data must be complete enough that the buyer can move the technology forward without the assistance of the seller.

This form of transfer generally works well when the seller has no interest in or resources for developing the technology further, or when the buyer wants to dominate the technology and all of its applications. A straight sale does not work for high-tech products that require consultation prior to purchase, extensive training in the use of the technology or continuing support from the seller after the transfer.

Publication

A publication transfer is the mechanism used most extensively by buyers and sellers. It occurs when the owner of the technology makes technology available in a public manner by writing a book, handbook or articles for a technical journal, or by providing educational services. Publication is the principal objective of educational institutions that exist to disseminate knowledge as widely as possible.

The principle of publication is the basis for the legal rights inherent in the patent and copyright: in order to advance science, laws were passed to reward inventors if they promised to "show all." Their reward is a government-backed monopoly which gives the rightsholder an exclusive right to commercialize his work.

But publication has drawbacks, too. It is a form of communication in which a single speaker presents information to a large group, up to and including the entire world. There is a presumption that the public "listens" to the knowledge, and if it listens, that it understands the information.

This presumption works well in formal educational settings, but tends not to be true of larger forums for several reasons: the listener may not know if or when the speaker will speak; the information may be impossible to understand; and publication may get lost in the overwhelming tide of daily information flow.

In fact, this publication problem is the single greatest complaint against the dissemination of information on government research by the federal government. Eventually, these publication difficulties led to the decentralization of federal technology transfer activities with the passage of the Technology Transfer Act of 1986 (See Appendix B for more information about technology legislation).

Initially, the federal government tried to create a centralized clearinghouse called the National Technical Information Service (NTIS). Although it was a significant improvement, this agency proved inadequate because it relied on incomplete information which was provided by federal laboratories. It also lacked the necessary resources and staffing to interpret the information for prospective buyers. As a result, the Technology Transfer Act of 1986 shifted the responsibility to the federal laboratories, giving individual lab directors the authority to deal directly with the private sector. The NTIS remains as one method for identifying available federal technologies. In the meantime, however, the federal government formally recognized and funded an agency called the Federal Laboratory Consortium. This agency also serves as a central site for locating available technologies and works in cooperation with the NTIS and federal laboratories.

License

This is the most popular form of technology transfer because it transfers less than total rights in the technology, which appeals to most sellers. In effect, the technology seller divides the technology into separate parts, each of which represents a different opportunity based on either applications or markets. The seller can select any of these aspects as the subject of the transfer and retain the rights to other applications and/or markets.

The seller typically transfers only those rights which enable the buyer to capture the opportunity designated in the transfer agreement. The rights can be in the form of a patent, copyright, trade secret or a combination. But in effect, the seller transfers only enough rights to complete the negotiated transfer.

In a license transaction, the seller must come to the table ready to identify and describe that application or market he desires to license. Standing alone, the technology can serve numerous markets; therefore, the delineation of the subject market requires careful study and most of the time involved in pre-transfer planning.

On the other hand, the buyer must assure himself that he is obtaining sufficient rights to commercialize the technology within the given market. He also must verify that the grant of rights does not conflict with other rights already granted or that it is not diluted by a subsequent grant.

It is important to know that this form of technology transfer gives the greatest opportunity to maximize revenue from the technology because each license represents a separate market that can be valued independently at its maximum potential. For example, the seller may develop a photovoltaic cell which can be used to power industrial electrical devices, to heat residential homes or to provide straight power generation.

In this case, the seller can license each of the above functions of the photovoltaic cell separately to three different buyers. This results in the creation of three revenue streams, one from each of the buyers, which results in more profit potential for the seller than if he had licensed all three functions of the cell to a public utility for power generation alone.

The disadvantage inherent in a license transfer is that, although each license represents a potential profit center to the seller, it also requires separate and additional resources. As a consequence, the larger the number of licenses created by the seller, the greater the total cost to transfer.

If the seller cannot afford this sophisticated process, he can fall back on licenses that group different markets and applications. Most sellers opt to license when they have more than one defined market for their technology, which can be licensed separately for greater profits.

Resale

This form of transfer requires three or more parties: the first party sells to the second party and the second sells to the third. The second party, who is considered a facilitator of the transfer and a focal point for transfer activities, acts as a technology broker or dealer who takes a profit from the service provided when the transfer is consummated.

A resale transfer can be as simple as two straight sales in which the technology changes hands but not character. In other situations, the re-seller may improve or add value to the technology to make it more salable. This improvement can be nothing more than properly packaging the technology, or an improvement can be created by a buyer who assumes responsibility for one or more stages in the technology life cycle.

A resale transfer is desirable because of the inherent advantage of utilizing experienced transfer personnel to complete the transaction, so it is most beneficial when neither the buyer nor the seller has the expertise to complete all the tasks that follow the transfer process. In addition, this transfer is promoted by the interaction between the buyer and seller, or broker or dealer. Without this interaction, the buyer and seller might never have met.

The disadvantage, however, is that it increases the cost of the transfer because the broker or dealer charges a fee of up to 50 percent of the transfer price. This rules out the small transfers that cannot justify that fee level.

Most resales occur when the seller lacks the financial resources to complete a transfer and turns to a third-party broker or dealer to facilitate the transaction.

Cooperative R&D

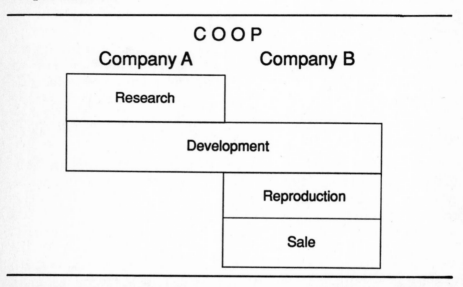

This form of transfer indicates two or more parties sharing the responsibility for all the research and development stages of the technology life cycle. It was popularized by the Competitiveness Act of 1984, which deleted registered research cooperatives from government antitrust actions, even in those cases in which a clear monopoly was established. Recent legislative proposals have sought to extend this protection to development activities, and to eliminate the private right of action for antitrust.

Cooperatives, which involve one or more entities, begin with the presentation of a concept by a single party to other parties who are encouraged to participate in the research and development phases of the life cycle. When research and development is completed, ownership of the technology can reside with someone other than the party who presented the original concept.

Cooperatives present an opportunity to distribute research and development risks among more than one party. In this way, participants gain by reducing costs: if the project is a bust, losses are limited; on the upside, if the participants in the cooperative are not direct competitors representing different horizontal or vertical niches, there is no earnings dilution. As a result, there is an opportunity to reap the benefits of the technology with a small investment as if the party had borne all the research and development costs. Most cooperative R&D transfers occur in the early stages of development, i.e., at the research or development stage when there is a substantial risk that the technology will not prove to be scientifically or economically sound.

Other cooperative R&D transfers are created when the seller's cost to complete research and development are so extraordinarily high that he wants to spread risk and/or cost over a larger group of deep-pocketed buyers. (See the Cooperative R&D Agreement in Appendix C.)

Contract R&D

This form of transfer is popular among partnerships between educational facilities and the private sector. It often is used to avoid or dilute the otherwise obvious participation of an educational institute in commercial activities.

This is an important avoidance mechanism because educational institutions often resist engaging in direct technology sales on the theory that such practices corrupt the academic system and cause a deviation from the primary objective of the university: student education. They compare it to excessively high athletic salaries that create All-American professionals who graduate from college despite an inability to read or write.

The structure of this transfer requires the educational institution to advance a research and development project that is already in progress; the private sector participant foots the bill.

Unfortunately, the educational institution seldom receives compensation for its prior development efforts, which can represent the largest part of the value in the transaction. The operating myth here is that all value in the technology is derived directly from the work funded by and commenced after the research and development contract is signed.

Loaned Servant

A "loaned servant" transfer occurs when there is an exchange of personnel involved in the development of the technology rather than an exchange of the technology itself. It is a desirable form of transfer when the transaction requires a lot of hand-holding to get the deal done.

In effect, one or more key individuals in the development of the technology decides to change affiliation; departing the inventor who originally brought the technology forward, this researcher or group follows the technology in its current form and continues work on the project—but for the buyer instead of the seller.

The loaned servant transfer of personnel can be temporary or permanent, contingent on the terms of the transfer contract, and the needs of both buyer and seller. Legal rights to the technology are passed from buyer to seller under the same terms as allowed in an ordinary sale.

The loaned servant transfer usually occurs at an earlier stage in the life cycle of the technology when little of the technology's potential use or benefit has been put into practice in a form that can be adopted readily by a potential buyer or end-user.

One of the most common users of this form of transfer is the federal government. Government agencies and departments prefer the loaned servant agreement when U. S. national security interests or other aspects of the public welfare are deemed to be best served by a rapid and painless dissemination of technical information into the hands of one or more qualified buyers who can get the technology to the commercialization stage.

Merger

Another transfer approach is to acquire the corporate owner of the technology itself in the merger of respective buyer and seller interests. By the nature of this transfer, it is limited to corporate entities or to

Chapter Four

those instances when the rare privatization of a governmental agency occurs.

A merger transfer requires that no exchange of ownership rights occurs; instead, ownership rights are consolidated to simplify the transfer process by limiting it to the necessary tasks of internal reorganization.

A merger typically results in an exchange of stock between the participating entities. In a merger, this amounts to a tax- free exchange creating a surviving company that may or may not keep the existing name of one of the companies.

Alternative forms of a merger transfer in which cash accompanies the exchange of stock can create a taxable event. (The taxation of technology transfer is covered in more detail in Chapter 7).

A merger has one major advantage over other forms of transfer: it captures all of the elements of technology development in addition to all of the critical resources used to advance the technology to its current stage in the life cycle. As a result, the merger works best when the buyer needs the technical or management expertise of the inventor or development team and/or their facilities, reputation or equipment.

On balance, however, there are definite disadvantages inherent in a merger transfer. The technology seller brings all the excess baggage and problems from prior development stages, including unneeded or unproductive personnel, and tax losses that cannot be used to offset revenue, among other problems.

To offset a potential imbalance in the advantages and disadvantages, sophisticated entrepreneurs create a merger transfer when there is a clearly dominant entity in the transaction who is willing and able to dictate how the merged operation should be conducted in a way that is optimal for both sides.

In the absence of a dominant party, there can be costly redundancies in personnel and internal struggles for control of the merged entity. A typical example is a large corporation that buys a small company in order to acquire a unique technology which the small company has developed.

There are two reasons: (1) often the buyer's merger transfer costs are lower than if he had started up a separate entity by which to develop the technology from scratch, and (2) the larger corporation also can acquire the markets of the technology or small company in a more cost-efficient manner with a merger.

Spin-Off

One very common transfer scenario includes a technology owner who wants to sell his process for any number of good reasons that support prudent management objectives, but he finds it difficult to let go because the technology appears to be extremely valuable.

This environment sets the stage for the spin off form of transfer in which the rightsholder creates a new entity in order to exploit the technology without losing all its benefits.

In practice, this means that the entrepreneur has no further development responsibility for the success of the technology, but he can participate in the benefits that accrue from its commercialization, in the form of either a royalty arrangement or a share of equity.

In most spin-offs, the transfer of technology results in a sale of stock shares in the new corporate entity and the technology is treated as a capital contribution. The title and all rights to the technology pass to the new company. If the technology seller receives no cash in the transfer, a low value can be placed on the technology and on the stock received for it so that potentially negative cash flows can be prevented.

If the seller takes cash and/or royalties from the new company, it looks like a standard transfer except that the new company retains important ties to the old entity. In either case, it is advantageous for the new company to develop its own objectives, operating plans and resources free of any restraints that might have existed in the old company.

The spin-off is considered extremely beneficial to commercializing new technologies that promise to develop more effectively off the beaten track developed by the former company, for example, when a food manufacturer develops a new device for quality control which does not fit existing food product lines but has applicability to the company's industry. Rather than diverting capital and other resources from the strategic goals of the company to promote its food lines, the company will create a new company in which to develop, manufacture and market the quality control product. This development often occurs within companies that need to develop instrumentation in order to further develop their existing products. In fact, the development of the equipment necessary to manufacture the primary technology can supercede existing products in potential use and profits.

This transfer mechanism recognizes that any attempt to commercialize the technology within the old company either could interfere with existing operations or could result in the failure of the technology in the absence of effective support.

But the spin-off concept can be carried too far by ambitious entrepreneurs: after the discovery of a new technology, there can be an irresistible temptation to create a new company or subsidiary for each potential technology application or market. If growth is not guided prudently, there usually are overwhelming, concurrent demands for additional resources from each of the new entities, which creates destructive competition for the parent company.

Reverse Engineering

Another common method for acquiring technology is reverse engineering, in which knowledge of a process is gained by studying an existing technology product to determine how it was made in order to recreate a competing product.

Reverse engineering is prohibited only by the existence of patent rights; reverse engineering can be used against a seller when his technology is protected by copyright or trade-secret rights because these forms of legal protection are inadequate to prevent anyone from using reverse engineering to gain more knowledge about his technology.

It is important to note that a copyright does not protect the ideas underlying the technology; this means anyone can use those underlying concepts as long as use does not constitute copying the copyrighted work.

The premise of a trade secret is, of course, secrecy. If the secret aspects of the technology can be determined by a study of the end-product, then secrecy is lost. Reverse engineering cannot be used when the technology is protected by patent because the patent grants a use monopoly to the rightsholder.

But reverse engineering can be used to determine how to go around the patent by using an alternative technology to provide the same or a similar benefit.

It is crucial to the technology entrepreneur to consider the inherent threat of reverse engineering when planning the commercialization of his technology. Those technologies most vulnerable to this threat, therefore, should be patented whenever that is feasible.

If patent protection is not feasible, the commercialization of the technology often becomes a dramatic race to the marketplace among worthy—and unworthy—competitors. In fact, the technology holder should go beyond protection to establish a firm, if not dominating, market share when the certainty of reverse engineering competition exists. When reverse engineering competition exists, the need for speed virtually dictates the acquisition of technology transfer partners who are well-funded and market-dominant in order to complete the commercialization stage.

Prospective technology buyers consider reverse engineering as a cheap alternative to purchasing or acquiring a technology in this form at a cost that is lower than standard royalty rates for the industry.

Following are the criteria by which reverse engineering is selected over other forms of transfer that would have cost the buyer royalties to the seller if the buyer had used a different form of transfer:

Reverse Engineering Checklist

- Can the technology be replicated?
- How much will it cost to replicate?
- How much time will it take to replicate?
- Does an existing patent bar replication?
- What is the demanded royalty rate if the technology is acquired by license or other form of transfer?

Chapter Four

Theft

Theft is the last form of transfer that should be given serious consideration by an entrepreneur engaged in technology commercialization. Theft, which is any unauthorized transfer of the holder's legal rights to the technology, occurs only when legal rights in the technology have been established prior to the theft. (A more detailed description of legal rights is found in Chapter 8.) It is vitally important to understand that unless legal rights have been established, there is no prohibition against the use of the technology.

For many years, government agencies and academic facilities promoted the free dissemination of technology developed under their collective aegis through publication. To the extent that this practice continues, theft is not an issue.

But the theft of similar intellectual property rights in the private sector is considered infringement. It can occur willfully, with full knowledge of the illegal nature of the act, or it can occur innocently. In willful infringement, the guilty party is considered to be aware that prior legal ownership was established in the form of a patent, copyright or trade secret.

The act becomes theft when the thief receives a benefit that is reserved to the legal rightsholder. Infringement, therefore, is illegal because the legal owner is harmed by an act that devalues his control of the technology and fails to reward him for his investment in the technology.

When infringement is innocent, the guilty party is considered to have acted without knowledge of the owner's legal rights. This occurs most often when a technology developer uses a patent invention after a monopoly has been awarded by patent to the original technology developer/owner. The infringer may have created the invention independently of the original creator—without access or reference to the work of the owner—but his efforts produced results later in time.

It is equally important to know that the onus of establishing legal rights, and enforcing them, is on the owner of the technology. His failure to respond to an infringer allows the infringement to continue, and can result in the complete or partial waiver of any legal rights that have been established on his behalf.

The caveat for all technology users is, first, to become knowledgeable about the potential existence of the legal rights of other technology developers which potentially could bar the use of a technology. The law contains an implied obligation on the part of entrepreneurs to seek out the prior existence of these rights through a search of registered rights.

A technology user will be unable to claim innocence in a patent infringement suit when information about existing legal rights is available to the general public for examination.

Because theft can be such a serious threat to the owner of a technology, it is sound business practice for most entrepreneurs to install programs that protect the technology and ensure continuous control for the rights holder.

Despite the fact that some technologies are much more difficult to protect than others against the guilty and the innocent alike—it is possible for some technology thefts to go undetected—even the innocent unauthorized use of technology subjects the user to the threat of litigation that can result in the assessment of damages, penalties and imprisonment.

5
Planning/Management Skills

The number of companies that transfer technology will double by 1992.

By itself, the management of technology is considered a very difficult and complex task. For this and other reasons, many entrepreneurs and others on the periphery of transfer activity label the process "unmanageable." One reason is that each transfer is considered unique, a separate and distinct event, having little or nothing in common with other transfers. However without denying the complexity of transfer and recognizing the great diversity of transfer mechanisms, transfer should be considered no less subject to management technique than any other business activity.

In the same way that almost any business activity or process can benefit from the implementation of management practices that reflect the real world, technology transfer also responds favorably to goal orientation that enables participants not only to increase the likelihood of success, but also to increase potential rewards.

A description of sound business practice rarely presents surprising information because experience teaches that good management has certain commonalities whether it is applied to a company, or to a technological process or to a transfer mechanism.

The first is setting objectives. The second is setting strategies that achieve the objectives. The third is determining the resources required to carry out the strategies. The fourth is selecting available strategies for the development of a business plan. And the fifth is creating an implementation plan to operate the business effectively.

When these management techniques are applied to technology transfer, however, what happens frequently is that sound practice is disparaged as a hit or miss process such as gambling. It has been said that almost any business venture is a gamble; without a doubt, the experts agree that technology management is even a greater risk than most manufacturing and service business operations.

Project Management Issues

- Transfer Management Should Emphasize Communication Among All Participants
- Informal, Face-To-Face Communication Is a Key Process in R&D Success
- Key Management Roles Are: Gatekeeper, Champion, Boundary Spanner and Sponsor
- All Transfer Participant Roles Are Both Formal/Legitimated and Informal/Spontaneous

---

Although optimal management can control a technology enterprise that merits commercialization, even a superstar technology that venture capitalists covet become a failed gamble without effective management control.

One example is the mid-1980s failure of Denver-based Storage Technology Corporation, which had developed computer storage devices. The company tried to diversify into several markets concurrently, pouring more than $100 million of venture, R&D and other forms of capital into the business before the company filed bankruptcy, a marketplace failure attributed to excessive cost overruns and lack of management focus on appropriate target markets.

Technology Transfer as an Objective

It is important for technology entrepreneurs to determine early on whether or not transfer of the technology is the primary objective. If it is, then a successful transfer concludes the project and is not meant to serve as a stepping-stone to a subsequent objective. Entrepreneurs should make the transfer determination early on because technology transfer as the primary objective impacts several other aspects of management in an important way.

For example, many companies in the private sector treat technology transfer as a business in itself. Technology dealers view the technology as a product; brokers, on the other hand, view technology transfer as a service. This applies whether the private-sector company is a for-profit or not-for-profit enterprise.

In the 1990s, it is projected that small-business owners also will experiment more and more with technology transfer as a corporate objective. Until this decade, the lack of a common marketplace for technology has limited transfer-as-business experimentation; but new global markets and the proliferation of companies of all sizes ready to transact some form of technology transfer promise to increase the revenue potential of technology transfer dramatically.

For this reason, technology "dealerships" and brokers look like the most promising facilitating structure for transfer in the coming decade (excluding attorneys). The reason is that brokerages will benefit from the entrepreneurial trend toward using an intermediary to effect a transfer rather than transacting the deal without outside assistance.

This trend indicates a marked departure from historical and current practice, which promotes do-it-yourself deals. In fact, it is estimated that only 2 percent of all transfers to date have been completed with the assistance of third-party professionals.

One of the clear advantages to structuring a dealership is the capacity for greater control over the myriad variables that a technology transfer entails; the most important aspects of the deal orchestrated by a dealer are the presentation and negotiation stages of the transfer.

The dealership, however, also is vulnerable to greater risk because substantially greater resources are required, first to acquire rights in

Industrial Outreach

GENERAL APPROACH: A more highly-structured, synergistic effort to promote interest and participation throughout U.S. industry

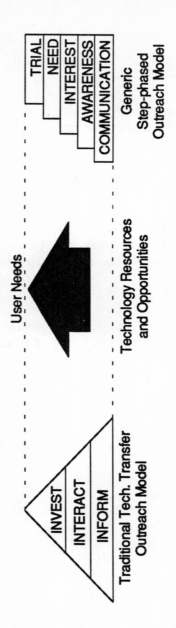

Source: NASA

The trend is toward a more sophisticated method of engaging in technology transfer because general publication of scientific discoveries is no longer recognized as a practical and effective way to bring a buyer and seller together. Today's approach demands that the entrepreneur go beyond networking and practice skills common to "niche marketing." Not only do transfer skills improve in niche marketing, but a marketplace begins to emerge with recognized buyers, sellers and facilitators.

the technology, and then to improve it to a point at which transfer can be completed successfully and profitably.

In the public sector, technology transfer has made increasing gains as an important objective in and of itself. Although it still plays a secondary role in the recovery of federal research dollars, it has become primary in global economic competition. The preeminence of the United States in science is viewed by many industry-watchers as a critical key to national security.

Since 1980, two significant pieces of federal legislation have been passed. The Stevenson-Wydler Act (Public Law 96-480) made technology transfer a part of each federal laboratory's mission. In turn, the Bayh-Dole Act (Public Law 96-517) allowed small businesses and nonprofits to claim title to any technology developed under a federal government contract. These laws were expanded and strengthened by the Federal Technology Transfer Act of 1986 (Public Law 99-502).

During this time, some federal laboratories—particularly those under private contractor operation—made technology transfer a primary part of their mission. Rather than concentrating on pure research, laboratories such as the National Aeronautical and Space Administration (NASA) and the Solar Energy Research Institute (SERI) expended research dollars with the intent of creating new technologies and industries.

Management Infrastructures in Academia

- Discipline-Based Department
- Organized Research Units (ORU)
- Guild Structure (Associations)
- Invisible Colleges (Information and Domain-Focused)

Source: The Industrial Technology Institute

In the academic sector, technology transfer always has been a principal, if not the primary, objective. But academicians rarely exercise effective control over their technology transfer—often control is deliberately shunned because of staunch academic principles that support only the free flow of information—so many of the benefits and much of the profit from technology transfer are not realized.

Corporate observers still question whether technology transfer is recognized as a primary objective for research-oriented universities, or whether it continues to function as a pawn in the ongoing political struggle of financing academic and research pursuits.

Before determining the objectives of a management plan to commercialize technology—whether it is from the private, public or university sector—several factors should be reviewed in detail by the entrepreneur. These are indicated in the outline below, which serves as a guideline for the thought process that must precede the development of a management plan.

Before Writing the Plan: Factors to Consider in Structuring a Cooperative R&D Agreement

- Authority
  — Technology Transfer Act of 1986
    (Pub. L. 99-502, amending 15 U.S.C. 3701, *et. seq.*)
  — 42 U.S.C. 203(c)(5)&(6)
    (NASA only, Space Act Agreements)
  — Other statutory authority
- Not intended to be sole or exclusive authority:
  "This authority is optional—and is not intended to affect previously existing cooperative agreement authority, such as the Space Act provisions, which for almost three decades have permitted NASA laboratories to enter into cooperative agreements."
- Subject matter of agreement
  — Use of facilities and/or equipment

Chapter Five

- — Sharing of information, technology or know-how
- — Sponsored R&D activities
- — Joint conduct of R&D activities
- — Any combination of above
- Statement of responsibilities
  Who is going to do what, when and where
- Funding arrangements
  - — Each party funds own activities (cooperative agreement)
  - — Private party reimburses government for its activities
  - — 15 U.S.C. 3710a precludes government funding of private party
- Intellectual property rights considerations
  - — Made or produced by private party
  - — Made or produced by federal employee
  - — Made or produced jointly
  - — Should be tailored for each individual type of agreement, depending on subject matter of agreement, the statement of responsibilities and funding arrangements.
  - — Should be limited to rights arising out of activities performed under the agreement; background rights should not be required to be introduced into the agreement.
- Inventions made by private party
  - — If the activity is fully reimbursed by the private party, the government should acquire no rights (at least under Space Act agreement). 15 U.S.C. 3710a appears to require government to receive government-wide license rights whenever it waives any right of ownership. It is unclear as to what is this right being waived.
  - — If the activity is of a cooperative nature, either: there is no exchange of rights or the government may require a royalty-free, nonexclusive license for governmental purposes in exchange for a similar license to a private party.
- Inventions made by federal employees
  - — Can apply only to those inventions for which government has, or can, acquire rights under E.O. 10096.
  - — If the private party fully reimburses the government, it should either assign to, or give the private party the first option to

obtain a royalty-free exclusive license subject possibly to license rights in the government.
- If made under a cooperative agreement, consider either:
  — No exchange of rights
  — Exchange of non-exclusive license rights only
  — Exchange of nonexclusive license rights coupled with a first option in private party to obtain a partially exclusive, royalty-bearing license (terms and conditions to be negotiated)
  — Exact approach should depend on combination of commercialization objectives and possible government use
- Inventions made jointly
  — Rights to the government's individual interest, again, is subject to those the government has or can acquire under E.O. 10096 and 37 CFR Part 501.
- Data produced by private party
  — If the activity is fully reimbursed by a private party, the government should acquire no rights to the data.
  — If the activity is of a cooperative nature, it may exchange data with unlimited rights, but the private party should have the opportunity to retain rights to detailed designing, manufacturing or process data pertaining to an item, component or process the private party intends to commercialize.
  — Government would have government purpose rights subject to nondisclosure prohibitions only
  — May consider time limitations on the commercial rights
  — May consider some type of march-in rights
  — If it is necessary for the private party to introduce background data, it should be fully protected, and used and disclosed only to the extent necessary for the government to fulfill its responsibilities under the agreement.
- Data produced by federal employees
  — All such data is at risk under the Freedom of Information Act
  — If the activity is fully reimbursed by the private party, the government should acquire no rights to the data, and promptly return it to the private party or agree not to disclose or publish it until reviewed by the private party, to the extent permitted by law.

- Protection will be strengthened when the data in question is closely related to the private party's proprietary data.
- If the activity is of a cooperative nature, the government may use its best efforts to the extent permitted by law, to withhold from disclosure detailed design, manufacturing or process data that may undercut commercial rights to any item, component or process the private party intends to commercialize.
- Protection may be strengthened if this data is closely related to that which the private sponsor has produced for the same item, component or process such that the release of either would cause commercial harm.
- In any event, the government may withhold from disclosure any information that may disclose an invention for a reasonable time to the extent permitted under 35 U.S.C. 205.

- Other legal considerations
    - Limitations on resources to be committed by the government
    - Representations, warranties and disclaimers
    - Mutual observation of rules
    - Title to property and waiver of damages
    - Limitations on liability
    - Survival of rights
    - Applicable law
    - Officials not to benefit
    - Availability of appropriated funds

Source: Robert F. Kempf

Technology Transfer as a Strategy

When technology transfer is not an objective, it is often an important strategy by which to accomplish an objective, whether the objective is making a profit, promoting education or providing for the common welfare of a nation. In some instances, it is the best strategy for achieving the objective. In others, it is the only strategy. The entrepreneur must consider technology transfer relative to his corporate resources, as well as to the inherent strength of the technology itself;

transfer can be used to yield better returns or simply to avoid failure.

For example, sunglasses offer such high profit margins that they can be offered as a single product line in proprietary stores. Compare this product with the profit margin on auto parts, which have to be grouped in large numbers of products because each individual part offers low margins.

An example in which transfer is the only management strategy is when the technology is an improvement on an existing technology, and the existing technology is controlled by a third party.

The technology manager, therefore, should determine not only whether or not transfer will advance corporate objectives and plan accordingly; he also must recognize his own ability to self-commercialize the technology—or his lack of it. Several non-transfer, primary objectives that apply to the seller as well as the buyer support the use of technology transfer as a strategy for reaching those objectives, including:

- The seller or buyer prefers to limit his involvement to certain stages of the technology life cycle and not others. For example, an engineer may be proficient at building prototypes, but dislikes personal interaction and corporate administrative duties. Transfer allows him to profit from his participation only in the first two or three life cycle steps which do not require business acumen to the degree of later stages.
- The seller or buyer can achieve maximum efficiency by limiting his involvement to selected portions of the technology life cycle. For example, an inventor can develop two new products every year by limiting his involvement to the product development stages. If he were to complete all seven stages of the life cycle, he would be limited to the development of only one new product every year.
- The seller or buyer can maximize his rate of return by not participating in certain stages of the technology life cycle. For example, the cost of manufacturing allows only a 5 percent allocation of profit to the manufacturing step if he transfers at the manufacturing stage. Therefore, he may elect to subcontract out

the manufacturing step and create a rate of return on the other tasks that could be 10 or even 15 percent.

- The seller can leverage his resources by shifting development responsibility to another entity. For example, if a product requires an investment of $50 million in tooling and plant construction to complete the manufacturing stage, a developer can contract with another entity to perform that task and use the $50 million for the performance of another task.
- The buyer can improve his net profit per product by distributing his overhead over a larger product line. If the buyer has an overhead factor of $10 million and generates $12 million in revenue, the acquisition of another technology can cause only a nominal increase in overhead and yet make a significant impact on his revenue.
- The buyer can decrease his risk of investment in research and development by acquiring an existing technology. If the scientific and economic proofs have been completed, no buyer R&D risk remains.
- The buyer immediately can meet or beat competition by acquiring rights to the best available technology.
- The buyer can decrease his research and development cost by sharing responsibility with other entities.

Technology transfer is the only viable strategy when the rightsholder does not have the necessary capital and resources with which to complete self-commercialization. Technology transfer also is the best strategy when a manufacturer, distributor or retailer cannot afford the vulnerability to risk during the research and development stage. This means the buyer opts for a totally risk-averse position and does not participate in R&D activities. So all new products must be acquired to avoid R&D risk.

Resources

All corporate strategies require certain resources in the form of capital, facilities, time, expertise and equipment. However, most entrepreneurs fail to consider the cost of transfer until all available resources have been exhausted. The following is a review of corporate resources that must be managed effectively prior to transfer.

Capital

Although most financial and corporate experts believe that technology entrepreneurs "cry wolf" when they claim there is never enough capital to commercialize, the highly-variable cost of commercializing a technology usually exceeds what the undercapitalized entrepreneur can spend.

A conservative capital guideline is that most entrepreneurs discover they need to spend at least twice what they had budgeted for the commercialization process—much like the venture capital business plan review process, which demands that new owners double expenses and halve potential revenues to achieve a realistic picture.

As outlined in Chapter 3, the process of transferring technology involves many expensive tasks that the seller must assume he will be required to finance, although it is possible to share some of the tasks between the seller and the buyer. So the optimum time for the seller to transfer to avoid overwhelming resource demands is when the research and development stages have been completed.

A classical transfer usually occurs after commencement of the research stage and prior to the manufacturing stage. During this time period, the technology is still incomplete but the proofs have been completed and high manufacturing expenses have not yet been incurred.

After financing the research and development stages that yield the scientific and economic proofs, the seller must carry the financial burden of locating and contacting prospective buyers, preparing and delivering a presentation, and initiating extended legal negotiations.

A good rule of thumb is that the cost of technology transfer through the proof stage does not vary significantly from the cost of raising venture capital as described above. In fact, if technology developers complete the scientific and economic proofs, most would find the venture capital solicitation a much less painful process.

But venture financing is considered an expensive way to finance a business, both because of the high returns guaranteed to the backers—at least 30 percent compounded returns per annum and five to seven times initial investment in three to five years—and because of the many hidden costs in solicitation time and additional management overhead demanded by venture capital firms.

Technology transfer also contains many unplanned, out-of-pocket expenses, including the expense of domestic and international travel to seek scarce financing and to sift out undesirable transfer partners.

Personnel

Like most other corporate strategies, certain requisite skills must be in place to complete a transfer successfully. Technology transfer draws on a diverse selection of talents, including skillful and experienced management, the financial expertise of bankers and venture capitalists, the legal background to address intellectual property issues, the engineering experience to handle technical issues, the accounting knowledge to avoid complex taxation problems, the marketing know-how to leverage user and industry profiles, and the sales skill to present and move the end-product.

Because only a minute percentage of the American entrepreneurial population has all of these skills, technology transfer must depend on effective team-building at the management level.

Like the aggregate of skills necessary for successful technology development, the relationship of those managers to the company is a significant factor in the success of the commercialization effort. When transfer is a primary objective and the technology is transferred in its entirety, the management team generally is dissolved when the transfer is completed. When transfer is a secondary strategy, the entrepreneur cannot afford to idle the management team until a transfer occurs. Even under the most optimal timing and management of resources, the cost of assembling managers and service providers at market rates can absorb any or all potential profits from the transfer.

When that realization dawns, many entrepreneurs react by assembling a less qualified team in order to save money. But that strategy backfires because under-qualified teams seldom have the expertise to complete transfer deals.

The last key player on the management team should be a sociology expert. More often than not, one of the hidden reasons a transfer defaults is the company's failure to address cultural differences between government and the private sector, between academia and the private sector, or between academia and the public sector, contingent on the kind of transfer involved.

At least one manager who can understand and appreciate the objectives, values and methods of the buyer—particularly a foreign buyer—is exceedingly helpful in almost any transfer deal.

Technology

The entrepreneur's technology—considered the requisite resource in the transfer—can not be evaluated realistically unless the scientific and economic proofs are in place. Because concept-stage technology is virtually impossible to transfer profitably, any attempt to forge ahead with a premature transfer usually constitutes a quasi-deal based on opportunity, i.e., a get-rich-quick scheme.

The best way to enhance the value of the technology is to demonstrate the newly-developed process in the form of a simulation, model or prototype, which can cost up to 30 times more than manufacturing a finished product in large quantities.

Time

The last, and ultimately most important, resource is time: time to find a buyer, time to put an advantageous deal together and time to complete the transfer correctly. Seasoned entrepreneurs know that technology transfer is not an instantaneous event.

Even when the resources, capital, environment and market conditions are positive, the transfer of knowledge takes more time than the entrepreneur can anticipate realistically in the planning process. (See the transfer timeline in Chapter 4 for an indication of how long each stage in the transfer can take.)

The key to a successful transfer at this point is a commitment by the seller to spend enough time on the transfer to satisfy all the requirements of the buyer. All too often, the seller turns to more lucrative, more interesting projects immediately after the buyer is contacted initially so that he can continue his income stream.

Without the assurance that sufficient time will be invested in the deal by the seller, the buyer will walk away nine times out of ten because he reads that situation as a lack of commitment. A partially transferred technology is worse than no technology at all because it puts the seller into red ink for no yield.

Although time is not typically thought of as a corporate resource, it is a crucial element of another aspect of technology management: many technologies fail the commercialization process as often from faulty timing as from not allocating enough time to complete the process correctly.

When faulty timing is in play, market introduction is either premature or tardy, which drastically affects the value of the technology. When competitors get there first, the user market is under-educated or world markets have not yet been created for the technology.

Therefore, the technology transfer management plan has to factor in enough time—as well as proper timing—in order to commercialize successfully. One example is the VCR, which was introduced prematurely and was relatively unprofitable until the development of cable television and the home video market supported its more widespread use.

Another example is the development of the beta-based video cassette, which was eclipsed soon after its introduction by the adoption of the VHS video cassette as the standard for the industry. Some industry devotees still remember the 8-track tape.

Planning

After the objectives and strategy have been determined, the best methodology for reaching the objectives must be developed. In technology transfer, the methodology or operating plan is identical in form to a business plan: it contains the mission statement of the backers, the objectives to be accomplished, the source and use of resources, and the tactics to implement the plan. The underlying elements of the operating plan for technology transfer are the assumption of risks and hedges. (See the Oak Ridge presentation in Chapter 6.)

Assumption of Risks

One of the most important elements of the operating plan is a strategy that addresses the level of risk deemed acceptable by management. No matter how many applications or markets for the technology exist, the existence of risk and the possibility of failure are always present. The risk level assumed by the entrepreneur should leave more than ample margin for error in terms of the time, capital and resources he can afford to lose.

The best plans adopt Murphy's law as a guideline and do not expect progress to go smoothly. Sophisticated financial reviewers say many operating plans and methodologies penalize technology development unnecessarily because they are so tightly constructed that even a minor obstacle can be a costly error in time and money.

The operating plan should include not only the life cycle of the technology, it also has to allow sufficient time for market acceptance of the technology. This means the management plan as an operating document should span five to seven years in order to factor in warranty claims, counter competitive technology and valuate the generation of significant revenue streams.

Hedges

No planning process has been executed so perfectly that it cannot be amended during the course of research and development. It is good business practice to select an alternative strategy in the planning process, and to select it in advance.

After a review of several strategies, it may be determined that the nature of the research does not allow a ready alternative or that the implementation of an alternative is difficult. In that case, it is more advantageous to select an original strategy that can be amended more easily.

For example, some strategies are an all-or-nothing approach which does not allow the developer to back out and start again. A developer may decide to complete the life cycle and not transfer the technology; but he realizes too late that he has run out of resources prior to completion and that he has effectively foreclosed his opportunity to transfer with a buyer who has found another outlet.

Or he may select a single application of the technology excluding the development of all other applications; too late, he discovers that he has selected a less desirable application and the opportunity to license has been lost.

Tactics

After an operating plan has been adopted, it should be infused with an overlay of "operating creativity" that will enhance the level of commercialization success. The implementation of operating creativity is particularly vital when industry standards for technology transfer are lacking in the entrepreneur's field of expertise. Many owners discover, to their surprise, that even the study of other technology transfer deals are of little benefit.

In operating creativity, tasks are staged so that performance and progress can be evaluated and reevaluated on a continuous basis. Sometimes referred to as "step-wise refinement," operational details outlined in the plan usually increase gradually as each development task is performed.

Staging allows a constant assessment of how successfully the plan is being executed, and it triggers various mechanisms for cancellation or modification of the plan early on if the plan proves unworkable. A definitive benchmark is determined for each development stage. Plans that are geared to unmeasurable goals provide no determination of task success until errors have been committed and resources have been wasted.

In other words, both time and monetary budgets should be established for each stage in the life cycle; these benchmarks should be met successfully before advancing to the next stage.

6
Marketing Skills

The marketing budget should allocate 10 percent to the identification of end-users, 10 percent to the identification of buyers, 20 percent to packaging and 60 percent to the presentation.

The unique aspects of technology as a product create equally unique marketing challenges that are either ameliorated or exacerbated by the level of technology completion at the time of transfer of ownership.

The earlier in the life cycle technology is sold, the more difficult it is to market; in many ways, marketing a technology transfer is analogous to the adoption process. The agency seeks qualified parents to assume responsibility for raising children just like an entrepreneur seeks an individual or corporate entity to assume responsibility for completing the commercialization of a technology.

The search for a corporate "parent"—including indepth research, telephone and face-to-face interviewing, due diligence investigation and assessment of the finalist candidates—requires the evaluation of several key marketing components as they apply to the potential participants shown below.

Benefits/Needs

There should be a direct and obvious relationship between the technology and the end-user of the technology through the benefits that address user needs. However, there may not be such an obvious relationship between the technology and the prospective buyer. The important difference is that the buyer's needs are not fulfilled by the use of the technology, but rather by the buyer's ability to commercial-

Technology Transfer Situations

Transfer Areas	Within the Same Company		From One or More Companies to Another	
Systems Involved	Isolated Individuals A	Groups B	Within a Country C	Within Different Countries D
Men (management)	Example: reception of a recent graduate, and reinsertion an experienced manager following intensive mangement training	Example: introduction of strategic management	Example: orienting a university research center to industry	Example: introducing American management methods in Europe (the 1950s and 60s)
Men and machines (production, maintenance, setting, research and development, data processing)	Example: orienting a surveyor to an instrument control setting section	Example: applying a research—discovered technology to manufacturing	Example: introducing builders to future users of new factories—NASA to American industry	Example: German space technology to the USA and Russia (around 1945)

Source: *Technology Transfer: A Realistic Approach,* Silvere Seurat, Gulf Publishing Company.

In the graph above, the transfer takes place first between the company and an employee and back again from the employee to the company in Situation A; between related groups within a company in Situation B; between the government and an individual/company, between two companies or universities, or between a university and an individual/company in situation C; and between two individuals, companies, universities or governments in two or more nations in Situation D.

ly exploit the technology to end-users. Typically, full exploitation occurs by selling the technology to the user.

Like products and services, technology often is marketed to elicit emotional purchasing decisions unrelated to the merit or use of the technology. When technology transfer marketing is based on hype—or blue-sky fantasies of enormous potential earnings—there is a tendency to push benefits to the end-user while failing to determine what, if any, buyer needs are met by the technology in its unfinished state.

So technology marketing—as opposed to technology transfer marketing—should focus on businesses that market to end-users and would benefit from selling the technology. Technology transfer marketing should identify those prospective buyers who will receive sufficient benefit to justify the purchase, in order to meet the corporate buyer's needs as well.

Identification of Technology End-Users

The second crucial element of marketing analysis is the identification of prospective technology end-users. A by-product of this task is finding a direction in which to look for prospective buyers of the unfinished technology.

After potential end-users are identified, projected technology benefits can be weighed against the user's need for valuation. If the technology does not wholly satisfy the end-user's needs, a value, if any, must be assigned to partial satisfaction.

This assessment often becomes highly theoretical rather than related to market value in the real world. For example, there are commercial air purifiers such as the ion generator which completely eliminate air odors and these have a marketplace value of $80 or more. Other products like aerosol pine sprays merely disguise existing odors—only a partial benefit—and, therefore, they have a value of only two or three dollars in the marketplace.

Another byproduct of identifying technology end-users is the identification of potential competition for the finished technology product. If the entrepreneur's product provides no more benefit than existing technology already provides in the marketplace, the most likely scenario is that existing market share will be subdivided even

Strategies For Promoting Technology Transfer to the Private Sector

	Demand—Pull Passive	Technology—Push	
		Role— Directed	Organization— Directed
Purpose:	To make information accessible to those individuals and organizations searching for solutions to customer/society problems	To actively promote awareness of new technology to individuals occupying boundary - spanning roles in organizations	To actively promote the adoption of new product or process concepts to innovator firms in an industry
Factors influencing selection:			
-Stage of R & D development	Early	Middle	Late
-Nature of innovation	Ideas or physical goods	Ideas or physical goods	Physical goods
-Available financial resources	Limited	Moderate	Extensive
-National priority	Long-term	Moderate term	Near-term
-Security concerns	No	Yes	Yes
-Willing to invest in assessment studies	No	Yes	Yes
-Willing to invest in concept studies	No	No	Yes
-Number and characteristics of firms in an industry	Diffuse	Diffuse	Focused
-Distribution channel design	Pull	Pull	Push
Technology transfer Strategy:	Technical databases	Professional journals and conference presentations targeted to certain disciplines	Personal contacts
	NTIS		Transfer of R & D personnel
	Professional journals	Trade publications and conference presentations targeted to certain industry groups or national associations	Onsite visits
	Trade publications		Joint ventures
	Conferences, workshops		Demonstration projects
		Technology fairs	Tax Incentives
		Industry teams	

Source: Richard O. Weijo, Pacific Northwest Laboratory, Richland, Washington

Because only five percent of more than 30,000 patents owned by the federal government are licensed for commercial use in the private sector, the federal government annually updates methods by which to transfer new technologies more effectively to the private sector. Sophisticated entrepreneurs, therefore, attempt to match their resources and inventions with the strategies used by government transfer policy makers, as shown in the graph above.

Chapter Six

Bargaining Ranges

Licensor	Licensee
Ceiling: Minimum of: (i) NPV of incremental profitability or cost saving (ii) NPV of cost to licensee 4of obtaining technology elsewhere	*Ceiling:* Minimum of: (i) NPV of licensees own R & D costs (ii) NPV of payments demanded by alternative supplier (iii) NPV of licensees estimate of incremental profit or cost coming from the technology (iv) NPV of costs of infringement.
Floor: NPV of transfer costs and opportunity costs of time and resources	*Floor:* Estimate of NPV of licensors transfer costs.

Negotiating Range

Source: *International Technology Licensing,* **Farok Contractor**

In price negotiation, the minimum and maximum prices are presented as perceived both by the buyer and the seller using net present value (NPV) of costs and payments.

further among the competitors. The existence of competitive technology also places a cap or ceiling on the allowable price that can be charged end-users for the technology.

A formula for establishing a price range based on the net present value of both costs and payments is explored on page 203.

Identification of Prospective Buyers

Identification by Capability

An alternative approach to the identification of a prospective buyer is to list the tasks remaining in the technology life cycle. For each task, the entrepreneur can seek a business capable of performing the task based on experience in the industry or product which, in effect, is a sub-contractor function. For example, if one of the tasks involves plastic molding, the entrepreneur would investigate various plastic job shops interested in performing the work under contract. In this situation, the shop is offered a piece of the action instead of a straight cash payment in exchange for completing the work. Sub-contractors should be solicited with a polished presentation in order to induce them to assume some of the technology's completion risk. Subcontractor prospects can be found in trade directories, in the Yellow Pages and through word-of-mouth.

Another way to identify potential buyer capabilities is to identify the seller's organizational "gatekeepers," as indicated in the next diagram, and assess the information they receive from the industry.

An alternative method of locating appropriate buyers is word-of-mouth; this method is used frequently by inexperienced entrepreneurs, but it is highly inefficient when exact credentials and expertise are sought. An initial, word-of-mouth inquiry can be opened within the specific field of science on which the technology is based, and there is some likelihood that it will yield an appropriate relationship to one or more industries that relate to the technology.

But sophisticated entrepreneurs prefer less subjective methods of finding a buyer: many approach prototype development companies with a proposal, i.e., those companies known to have had direct experience with later-stage or successful corporate entities in the entrepreneur's industry.

The Gatekeeper Network in a Geographically-Dispersed Organization

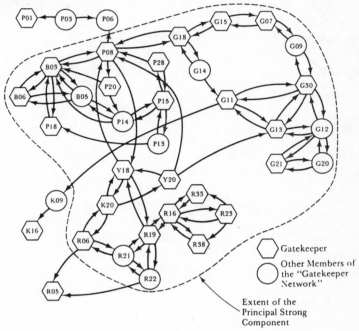

Source: *Managing the Flow of Technology: Technology Transfer and the Dissemination of Technological Information Within the R&D Organization,* Thomas J. Allen, The Massachusetts Institute of Technology

In every organization, there are one or more executives who have the decision-making authority to engage in technology transfer or at least suggest it to management. Finding these executives is a key step for the technology seller, and the search can be relatively easy if the manager has the title and position, for example, of new-product acquisition. It may be more difficult in other organizations, however, requiring further investigation among scientists or engineers at the target company during professional meetings and conferences, among consultants who were former employees of the company or among organizations of company presidents. In typical research environments, new information is brought into the organization through the gatekeeper. It then is disseminated to other gatekeepers through an informal network and sent outward from one or more points in the network to other members of the organization.

An important caveat in finding a buyer is their often surprising lack of interest, which is not uncommon. Many potential buyers contend with complex, later-stage management or cashflow problems and, as a result, are unable or unwilling to take on the risk and expense of outside technology.

Whatever buyer approach is deemed the best one contingent on the needs and resources of the seller, the most successful buyer "pitch" is a risk-free, sub-contract arrangement in which he receives a cash payment.

Identification by Distribution Channel

An even better way to find potential buyers utilizes the seller's knowledge of the distribution channels that serve the existing end-users of the technology. In this case, the potential buyer manufactures, distributes or sells a complementary product line.

When the seller's technology product is added to the buyer's existing product line, there can be significant secondary benefits in addition to profit realized by both the buyer and seller.

For example, when the seller's technology is distributed along with the buyer's product line, it can create additional demand for both products and/or open new markets that prefer the expanded line.

In addition to being an efficient way of finding credible buyers, identification of distribution channels is an easier approach. First, the distribution channels that would apply most effectively to the seller's technology product group is determined. Most trade directories identify by name the companies and products/services within a product group.

Another way to identify distribution channels is to survey prospective end-users to determine when, where, how and why they buy products similar to the seller's technology. The seller also can identify sales outlets, and product suppliers or manufacturers to get the same kind of information.

In either case, the second step is to assess each potential buyer's ability to commercialize the seller's technology.

Packaging

Packaging a technology translates all the data gathered and cross-referenced in the market research effort into a sales vehicle that supports the ultimate commercialization effort of the buyer.

The key to packaging this crucial information that accompanies the technology is the clarity with which it demonstrates the incentives to buying AND using the technology. So, by necessity, packaging must include both the scientific and economic proofs for the buyer, as well as the user benefits for the buyer's end-users.

Ideally, it answers in advance all the potential questions a buyer would ask in the due diligence investigation. In summary, it provides enough data for the buyer to become interested quickly and to believe that acquiring the technology will yield benefits.

The most effective demonstration of that message is the substance of the packaging: it actually demonstrates the scientific and economic proofs, including assurances, if not guarantees, that the technology works in real-world environments and that the buyer will make a profit.

If the prospective buyer acts as a sub-contractor, he may be concerned only with the potential profitability of the technology because a separate corporate entity will retain the liability if the technology fails to function as indicated in the packaging.

Successful packaging tends to be worded in the vernacular of the prospective buyer, who paradoxically may not be interested in, or fully conversant with, the scientific principles on which the technology is based.

If the buyer is a sub-contractor, he also may have difficulty measuring potential user benefits; as an intermediary, he sees his job primarily as a facilitator who simply moves the technology a little closer to the marketplace.

The bottom line in packaging is: prepare packaging that answers any and all questions the buyer may ask before he has to ask, in order to project an aura of credibility about the technology's ability to meet buyer—and end-user—needs.

Incentives

After potential buyer candidates have been identified, another level of market research must be conducted to determine what incentives lure these buyers into technology acquisition. The incentives are important because they can be used to identify additional candidates, to beef up the sales package and to empower transfer negotiations.

The best way to find out what incentives work for a given buyer is to get data that is unavailable to the potential buyer, or knowledge that the buyer does not possess. The best ammunition is an assessment of competing products and technologies, i.e., an assessment of the internal management of a competitive company's pricing or politics. The value of the incentives is determined by their value and importance to the prospective buyer, and it must be high enough that he is motivated to take action.

Another important facet of buyer incentives is how well they minimize risk. Buyers are every bit as risk-averse as sellers, and even more so when they face the unknown risks of commercializing the seller's technology. So the seller's presentation of incentives should quantify the inherent risks within the perspective of the buyer's ultimate profitability.

The Presentation

Any one factor in the commercialization chain can make or break a transfer, and that is nowhere more apparent than in the presentation of the technology. A superior, appropriate, prospective buyer can be targeted; his capability to perform the commercialization tasks can be unsurpassed; his need for the technology can be unqualified; he has become aware of the technology and is highly motivated to acquire it.

But transfer is not an automatic conclusion to a match among the seller's goods, the buyer's needs and end-user's benefits; the transfer often hinges only on how well the seller presents information at this stage in the process.

The presentation must be a potent combination of "sizzle" and substance that serves as the defense against any logical objections a buyer can raise. If the foundation is solid, the sizzle can make the sale. So it is prudent marketing to buy professional services to support the presentation, including well-designed graphics and clear copy that explains the technology in a lucid, vernacular style.

Some of the elements of a good sales presentation—including an exemplary market analysis and licensing projections—are illustrated in the transfer program plan between Martin Marietta Energy Systems, Inc. and the Department of Energy.

The Technology Program in Oak Ridge

The National Need

It is obvious that there is now a great national concern over the position of the U.S., and the future of the U.S., within the international economy. The stark reality of our international trade deficit suggests fundamental problems which extend far beyond the minor impact of unfair trade practices, if any, being employed by our trading allies. There is wide agreement that the successful economy of the future will be closely indexed to technology; not only the creation of technology of inherent value but, perhaps even more important, the ability to rapidly convert (apply) that technology to the needs of society in the form of product.

Within the U.S., about one-half of our total documented R&D expenditures are made by the U.S. government (roughly $59B out of $125B for CY-88). Most observers concede that very little commercial sector benefit has derived from these huge government expenditures on R&D. The reasons for this inability to successfully transfer technology from the point of origin, in this case the U.S. government, to the point of exploitation, the commercial sector, are manyfold; the historical barriers include statutory prohibitions, regulatory and procedural impediments, and even cultural and social taboos. The question has been: Can these barriers be lowered and, if so, would significant commercial benefit then derive from these large R&D expenditures of the U.S. government?

The Oak Ridge Approach

In 1983, when the Department of Energy was recompeting the contract to manage their vast Oak Ridge facilities (and the Paducah Gaseous Diffusion Plant), they requested all bidders to propose aggressive measures to transfer technology, particularly technology emanating from the prestigious Oak Ridge National Laboratory, to the commercial sector.

The successful bidder, the Martin Marietta Corporation, proposed a number of measures which they felt would permit the process of technology transfer to occur in greater measure. They further proposed, beyond simply permitting the process, to assume the responsibility to *cause* the Oak Ridge technology to be adopted by commercial firms.

The remainder of this paper is intended to summarize the program at work in Oak Ridge; it includes a description of the measures in place and the results of the program to date. This paper concludes with an assessment of the success of the program, what the DOE and Martin Marietta have learned from the program and what is indicated in terms of additional measures to expand and, perhaps, institutionalize the Oak Ridge program.

Measures Proposed

The contractor, Martin Marietta (Martin Marietta Energy Systems, Inc., a wholly owned subsidiary of Martin Marietta Corporation), proposed four primary measures, each of which, if adopted, would represent significant departures from the then current practices. The measures proposed were as follows:

1. To establish a new central function, the Office of Technology Applications, at the executive level with complete responsibility and authority to cause the transfer of technology.

2. To be granted an advanced waiver of the title to intellectual property, including patents, by the DOE. This was considered necessary as a means to give Energy Systems the authority and capability to license commercial firms in accordance with practices that are acceptable and attractive to the commercial sector. Another important feature of the Martin proposal was than any revenues resulting from the licensing of intellectual property rights would not be claimed as Martin Marietta profits but would be totally devoted to employee (inventor) rewards and the further enhancement of other technologies to the point of commercial attractiveness.

3. To implement an array of incentives, including financial incentives, for both the employees and the organizations at which the R&D takes place. Primarily, the financial portion of the incentives proposed were to come from royalty revenues generated from the licensing program proposed.

4. Establish, at Martin Marietta's private expense (investment), a "home" to accommodate technology client firms who may wish to locate in the area where the technology originated.

Measures Implemented

The Office of Technology Applications (OTA)

The OTA has been established, as proposed. This small group, headed by a vice president who reports to the senior vice president of Energy Systems, has the central responsibility for developing and successfully implementing the technology transfer program; including (1) evaluating the reservoir of available and evolving technologies, (2) selecting those with significant commercial value, (3) obtaining proper protection and rights to intellectual property involved, (4) identification and development of the commercial client firm, (5) the negotiation and placement of the technology licenses, and (6) the technical support of technology clients.

The Rights to Intellectual Property

An advanced patent waiver has been requested by Martin Marietta. Although the DOE has not yet been able to accomodate any general and automatic assignment of rights to Martin Marietta, both parties have agreed, in the interim of the approval of an advanced or class waiver of rights, that Martin Marietta will request the waiver of the rights on individual (case-by-case) basis. While the blanket waiver was considered to be the lynch-pin prerequisite of the proposed program, there has been, as will be shown later, significant success in the program even on the far less desirable basis of individual invention waiver requests.

The Evaluation of Evolving Technologies

Procedures have been developed and are now in place to systematically evaluate the commercial significance of all inventions. Now, for the first time, Energy Systems and the DOE have an improved basis for determining which inventions should go forward through the patent process and what markets (countries) in which to seek protection.

Licensing Practices

Martin Marietta has established, and has in operation, the expertise to select the preferred licensing clients, negotiate the licensing terms and formulate the licensing documents. None of this expertise previously existed within the contractor's organization.

Perhaps most crucial was the development of a standard licensing policy. Given the assignment of patient rights to Martin Marietta, they then have the authority to license in accordance with commercial practices. Martin can, for example, grant varying degrees of exclusivity if the circumstances so justify. . . . but when do the circumstances so justify? Which technology clients merit selection? When should the licensor impose performance commitments on the licensee and what is the nature of these commitments? The licensing policy attempts to address these thorny questions and add system and consistency to important licensing decisions.

Market Sector Analysis

If we are going to wisely license, say, a new superalloy, we need first to gain a familiarity with the superalloy market. How big is the market? Who are the corporate players and what are their market shares? The extent to which we inform ourselves on these matters will be the same extent to which we can develop and select the correct licensee(s) and negotiate the correct terms.

Again, this expertise has been developed and now resides, for the first time, in the contractor's (Martin Marietta's) organization.

Implement an Incentive and Reward Structure

The following measures, all new to the Oak Ridge operations, are now in practice at Energy Systems:

Chapter Six

1. *Inventor Royalty Participation*—Ten percent of all gross royalties received is distributed to the creators of the intellectual property which formed the basis of the revenue bearing licenses.

2. *Patent Application Award*—At the event of patent application, each of the named inventors receive a financial award of $500.00

3. *The Inventor's Forum*—The Inventor's Forum is a club of employee inventors managed by the inventors. Employees are admitted to membership in the Forum with the issue of their first patent. The Forum has programs, publishes a news bulletin, and presents each entering member a silver acorn pin (the pin is the symbol of the Forum—those with ten patents are presented a gold acorn pin). The Forum now numbers 600 plus members.

4. *The Annual Patent Luncheon*—All employees who have had a patent issued in the last year are treated to a luncheon once a year. The hosts are Martin Marietta, DOE, and the Inventors' Forum.

5. *Private Consulting Is Encouraged*—In the belief that technology best transfers with people, Martin Marietta has adopted, and DOE has approved, more liberal policies to permit and encourage private consulting between our technologists and our technology client firms.

6. *Entrepreneurial Leave-of-Absence*—Occasionally technology best transfers through the mechanism of new business formations or "spin-offs." Martin Marietta has instituted, and DOE has approved, procedures which encourage this process. (It should be noted that this practice, along with the practice of encouraging private consulting, are unconventional for private firms and have been put in place because of the Oak Ridge technology transfer program.)

7. *Institutional Rewards*—Most of the money derived from the licensing program will be used as a discretionary fund from which other technology maturation initiatives may be funded. This is viewed as an important form of institutional reward.

Accomodate New Business Start-Ups

As has been said previously, sometimes the highest confidence method for transferring promising new technologies is through new business formations dedicated to the single objective of exploiting one single technology. In these cases, one is balancing the usually higher risks of the new business being a success against the lower risks of the successful transfer of the technology. [The process of transferring technology is, in itself, not a simple process and not without risks. Locating a new business in the area where the technology originated, and where assistance is readily available, can lower these risks considerably. This is particularly true when the originating technologists are personally involved in the entrepreneurial endeavors.]

To accomodate this avenue of technology transfer, Martin Marietta established the Tennessee Innovation Center, Inc., (TIC). The TIC, a for profit subsidiary of Martin Marietta Corporation, has been established in Oak Ridge to meet the needs to foster and cause new business start-ups. The TIC offers a facility to accomodate the start-ups with quarters and a supportive environment, a staff to assist the new businesses with management expertise (to, hopefully, improve the "odds" of success), and seed capital for initial operating funds.

In return for the assistance noted above, the TIC negotiates a minority ownership (stock) position in the new firm. As some of these young firms may grow and prosper, the TIC may liquidate their equity positions and hope to both pay for firms that failed and still realize a profit on their total investment.

Obviously, there is nothing obligatory about new spin-offs, whether Martin Marietta employees or others, associating with the TIC in order to accomplish a transfer of technology. The TIC was established by Martin Marietta Corporation to meet an otherwise unmet need for yet another avenue for the transfer of technology. The TIC is managed directly by the parent firm of Martin Marietta Corporation and rigorous procedures have been adopted, and approved by DOE, to avoid actual and perceived conflicts of interest between the DOE's contractor, Energy Systems, and the parent firm, Martin Marietta Corporation.

Chapter Six

Accomodating Other Technology Client Firms

We have spoken earlier about the difficulty inherent in the process of transferring technology. Those acquainted with these difficulties express their conclusions in such phrases as: "Technology doesn't mail well," and "Technology transfer a contact sport." Said another way, co-location of the technology-originating organization and the technology-exploiting organization is, sometimes, a good idea.

To accomodate this notion, Martin Marietta established, at its own private investment, a modern industrial park in Oak Ridge, Tennessee.

Commerce Park in Oak Ridge, Tennessee, offers a well protected research park environment to technology client firms which may choose to locate in the area where the technology platform of interest already exists.

Results of the Oak Ridge Program to Date (1984–May 1989)

We will attempt to keep this section largely to a recitation of facts.

Intellectual Property Rights Assigned to Energy Systems

As has been previously described, no blanket or class waiver of intellectual property rights has been granted to Energy Systems. The preliminary technology evaluation efforts of Energy Systems indicates a portfolio of perhaps 60 families; separate licensing packages of technologies of significant commercial value presently exist. Of these, the rights to 21 technology families, comprised of 54 patents and 7 copyrights, have been assigned to Energy Systems.

- *Licenses Placed*
    - *Twenty-seven Commercial Firms Now Under Oak Ridge Revenue Bearing License*—The first patent rights were assigned to Energy Systems in August 1985. In the ensuing two years, 27 commercial firms have signed 30 licenses (see attachment).
    - *$635,000 License Fees Received by Energy Systems*—While a minor indication of eventual royalty potential, the receipts to date are still worth noting. It is thought that these same

already placed licenses may account for much larger royalty streams in the future.

It should be noted that the primary objective of the Oak Ridge licensing program is *not* the maximization of royalties but, instead, is to achieve the maximum benefit to U.S. economy via the creation of U.S. jobs and the causation of U.S. based capital investment. Nonetheless, royalty payments and future royalty prospects are important measures of the client's perception of value and, hence, worthy of notation.

$16 M of Commercial Sector Production Has Taken Place Under Oak Ridge License

While it usually takes two years (sometimes much longer) for a licensee to get a product to the marketplace, five of the Oak Ridge licensees have already paid running royalties to Energy Systems indicating final product sales. This early indication is extremely encouraging and bespeaks the prospect of a manyfold increase in U.S.-based production under these same licenses in future years.

Industrial Contracts and Visitations Have Multiplied

Industrial visitations to the Oak Ridge facilities have consistently increased since 1984. In 1987 there were 617 such visits, a 55 percent increase over the year before.

Employee's Requests to Privately Consult and Employer's Approval Have Increased Significantly

In the year 1983, there were 60 requests made and approved; in 1987, there were 129 requests made and approved.

Eleven New Business Spin–Offs From Oak Ridge Technologies In 1988

While it is not possible to construct accurate historical comparisons, we are certain that these activities are increasing at an impressive rate.

Chapter Six

The TIC Has 17 Client Companies

Of these 17 client companies, a few are struggling or simply meeting expectations. However, 12 are either doing very well or have very rosy prospects. One of the client companies, perhaps the stellar performer to date, is four years old, has had consecutive annual sales of $600,000, $1,750,000, $4,000,000 and $8,000,000. The last three years have been solidly profitable and this firm now employs 103 East Tennesseans. This same firm, in 1988, occupied their own stand-alone new facility, and again projects a large increase in sales in 1989!

Although the concept of the TIC appears, at this point, to be valid, there is one discordant note. Of the 17 client companies, a few are based upon DOE/Energy Systems generated technologies. This is not as was expected.

Both Energy Systems and DOE believe that the rigorous procedures adopted to prevent the perception of conflict-of-interest have appeared so ominous to the employees of Energy Systems that a "chilling" or discouraging effect has set in. Both parties, Energy Systems and the DOE, are restoring the ability of the TIC to more fully meet the original purpose of technology transfer of the DOE funded technologies.

Industry Is Now Asking for ORNL Contract Assistance

With the perception that the centers of government R&D are now more accessible to the commercial sector, private firms are requesting help from them, particularly form the ORNL. In 1987 ORNL performed $6M of R&D under contract from 55 private firms. This compares with $3.5M with 45 firms in 1985. While these are still comparatively small numbers, the favorable shift is considered significant.

Inventor Employees of Have Received $97,232 In Royalty Sharing

On December 7, 1987, $24,148 was distributed to seven inventor employees as the first ever such distribution. Since that time another $73,084 has been awarded. This practice has reinforced a very positive and supportive environment among the employees of Energy Systems regarding the technology transfer program.

$75,500 Has Been Distributed as Patent Application Awards

This represents a new practice.

Invention Disclosures Have Recently Increased, Reversing a Five-Year Decline

These are interesting facts (see attachments) because they are evidence that the entire program appears to have had the desired effect of positively motivating our technologists. [NOTE: These writers do not contend that recognition, or even reward, *causes* creativity. Creators will create (whether, it is said, "it is needed or not"). The possiblity remains, however, that should the "system" indicate that good technology is appreciated, people may be prompted to record what they would not otherwise bother to record. Further, people might be prompted, again should the "system" indicate that it would be appreciated, to complete the formulation of technological strategies that they might not otherwise complete.]

Technical Publications Have Increased, Again Reversing a Five-Year Decline

These facts (see attachment) are particularly interesting in that one of the classic arguments *against* royalty bearing licenses and any measures of direct employee rewards is that the employees would tend to become secretive and less willing to communicate with professional peers. This somewhat cynical argument appears to be debunked by the Oak Ridge experience, facts, to date. Interactions with others, both in publications and, as previously noted, in industrial interactions have dramatically *increased*.

Regional Economy Turns Up

Since the 1940's establishment of the Manhattan Project, the federal "reservation" and then the City of Oak Ridge has been considered an enclave of secrecy and exclusive federal purpose which offered little or no opportunity for economic dividends to the state and region beyond the payroll of the installations. Both the DOE and the community

Chapter Six

leadership viewed the 1983 recompetition as an opportunity to exhort the bidders to propose positive measures, including their own private investment commitments, to diversify the economy of the "one company" town of Oak Ridge. You have read earlier in this paper of some of the investment commitments offered and subsequently implemented, by the winning bidder. It is thought that these specific multimillion dollar commitments, while significant in themselves, pale in comparison to the potential of the economic dividends to the community of a truly effective technology transfer program. Although we cannot accurately trace cause and effect, let us look at some facts regarding the Oak Ridge economy during the last three years.

It should be noted that the data relates to a period in which the Oak Ridge economy, almost solely indexed to DOE funding, received word of three financially disastrous program cancellations—the Clinch River Breeder Reactor, the Gaseous Centrifuge Enrichment Program and the removal of the Oak Ridge Gaseous Diffusion Plant from an operational status. Each were programs of multibillion dollar potential. All were cancelled. What has happened to the Oak Ridge economy? It is surprising. Note the following facts:

1. Private sector capital investment was $36M in CY-85 setting an all time high... but it was over $100M in CY-86; it is expected to be about $150M in CY-89.

2. Over 30 corporations, new to Oak Ridge, have established operations within the city since July of 1987.

3. The Oak Ridge city schools have increased enrollment for each of the last three years reversing a *20-year decline.*

4. New multimillion dollar hotels and restaurants have come on line within the last two years. (In a town of 28,000 people, these are "big deals!")

5. Real estate is suddenly a seller's market. Two years ago, there were 600 plus homes on the market in Oak Ridge. Today there are 190 homes on the market with an apparent pent-up demand.

May We Extrapolate?

We believe the practices already in place in Oak Ridge will see the following technology transfer accomplishments by the year 1992: $100M/yr. of U.S. based commercial production under Oak Ridge licenses, 25 additional new business spinoffs of government-funded technologies, 100 active licensees, ten percent annual increase in commercial firm contracting actions with the Oak Ridge (DOE) enterprises, and three major industrial locations to the Oak Ridge region.

Whether the above numbers are viewed, by the reader, as optimistic or conservative, they are, approximations of the perceived potential of the program and they are measures, also, of the scale of commitment from the partnership of the operating contractor, Energy Systems, and the cognizant DOE organization (the Oak Ridge Operations).

Should the liberties of the operating contractor be increased, the commitments for successful transfer may increase significantly.

May 1989
William W. Carpenter
Source: Martin Marietta Energy Systems, Inc.

When Finding Buyers Is Difficult

When prospective buyers are difficult to locate because distribution channels that serve prospective end-users do not exist, some entrepreneurs issue a press release to industry-related publications advertising their desire to find interested parties.

The content of the release should provide enough information and be well-written so that the publication editor prints the release and readers will be motivated to respond. Typical publications that are willing to publish press releases include scientific journals like *Journal of the American Ceramic Society*, trade publications like *Computer World*, and entrepreneurial magazines like *INC*. It is considered common practice to submit press releases or ads to these print media in order to find technology buyers and investors.

Another way to find buyers is to advertise or obtain a listing in publications and databases about new technologies and technologies

Chapter Six

Technology Transfer Methods for Overcoming Barriers at Each Stage of the Adoption Process

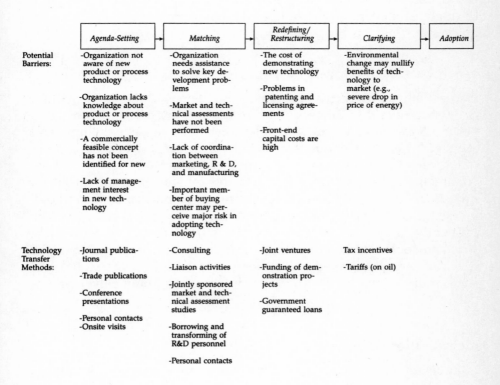

	Agenda-Setting	Matching	Redefining/Restructuring	Clarifying	Adoption
Potential Barriers:	-Organization not aware of new product or process technology -Organization lacks knowledge about product or process technology -A commercially feasible concept has not been identified for new -Lack of management interest in new technology	-Organization needs assistance to solve key development problems -Market and technical assessments have not been performed -Lack of coordination between marketing, R & D, and manufacturing -Important member of buying center may perceive major risk in adopting technology	-The cost of demonstrating new technology -Problems in patenting and licensing agreements -Front-end capital costs are high	-Environmental change may nullify benefits of technology to market (e.g., severe drop in price of energy)	
Technology Transfer Methods:	-Journal publications -Trade publications -Conference presentations -Personal contacts -Onsite visits	-Consulting -Liaison activities -Jointly sponsored market and technical assessment studies -Borrowing and transforming of R&D personnel -Personal contacts	-Joint ventures -Funding of demonstration projects -Government guaranteed loans	Tax incentives -Tariffs (on oil)	

Source: Richard O. Weijo, Pacific Northwest Laboratory, Richland, Washington

that are newly-available for acquisition. Many of these sources charge a fee to subscribers rather than charging the seller for his listing, for example, *Technology Catalyst* and *Technology Insights*.

A third alternative when a potential buyer is hard to find is the use of technology broker services. Brokers usually charge a flat fee or a commission based on a percentage of the monies received from the sale or license of the technology. Some examples of technology brokers are Arthur D. Little, the Big Eight accounting firm with offices throughout the United States, and Research Technology Corporation in Tucson.

When Qualifying Buyers Is Difficult

Occasionally, prospective buyers are extremely difficult to qualify. Even extensive due diligence efforts do not pay off, and continued effort constitutes an unacceptable waste of the seller's time and money. When that happens, it is helpful to get a list of the criteria other sellers have used to qualify their candidates.

Sometimes general criteria are available for similar companies from industry-related trade associations, brokerage houses or technology brokers. Not only do these guidelines help size up a candidate more efficiently, they also can help an inexperienced entrepreneur know when to look elsewhere. Dun & Bradstreet can supply a financial check on prospective buyers; a literature search in professional and trade publications often can reveal information about buyer principals.

Identification of Sellers

If the search for a buyer is conducted in reverse, the task of identifying sellers is both easier and harder than finding buyers. Government agencies and universities make buyer and seller identification easy because systems currently exist to facilitate technology transfer between the private and public sectors.

Although these facilities generally do not advertise their activities, a specific inquiry to any government department or agency, or to a university, can yield the information needed. A good source of technology sellers are the local branches of national economic development agencies such as the U.S. Economic Development Agency, or technology directories and databases such as Technical Insights, Inc. in Englewood, New Jersey.

In contrast, however, sellers in the private sector are much more difficult to identify—and the smaller they are, the more difficult they are to find—because often they are not grouped together in a common database of information or they do not belong to an industry-wide trade association.

7
Financial Skills

The total cost of a technology transfer—which ranges from $30,000 to $50,000 for an "average" transfer—usually exceeds the cost of completing the economic proof.

Technology transfer requires very specific financial resources because of the high level of costs incurred by both the buyer and the seller of the technology. In almost every case of technology commercialization, the success of the transfer is contingent on the availability and use of these resources.

Sophisticated entrepreneurs not only raise enough capital to develop the technology to the transfer point; they also are knowledgeable enough to factor in the additional high costs of the transfer itself.

All too often, sellers presume that their financial needs stop at the point of transfer; having exhausted their startup resources, they are unable to finance the completion of a transfer that would yield the financial payoff for all their initial investment and hard work.

The chart on page 126 indicates the major sources of financing for science and technology development in 1988 in terms of the distribution of state expenditures, including technology transfer:

Government Financing

One important source of technology and technology transfer funding for the seller or buyer is a state government program that is designed to increase the local tax base and create new jobs for local citizens. There are four major sources of technology transfer activity within state government, as seen on page 127.

"Science and Technology Initiatives"

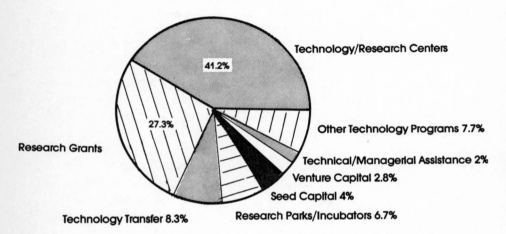

Source: Larry Crockett, University of Michigan

As indicated above, only 8.3 percent of the technology funding available at the state government level was allocated to technology transfer in 1988, a percentage which promises to increase significantly in the 1990s. To date, technology transfer funding is the third highest from this financing source, behind technology centers and research grant funds.

State Government Technology Transfer Activity

Guide: Graph

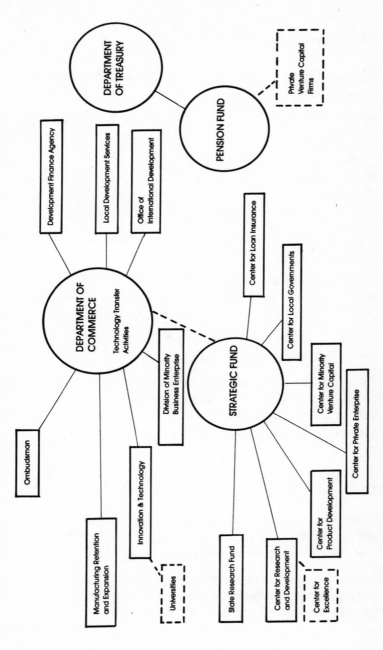

Source: Minnesota Office of Science and Technology

If a seller or buyer plans to solicit state government funding, there are a number of selection criteria which should be considered prior to submission of proposals, including:

1. State and local taxes per capita (indicates the relative burden of taxation in the state)

2. Percentage change in state and local taxes per capita (indicates the trend in the tax burden and the attitude of taxing authorities)

3. State spending growth versus state income growth (indicates the ability of the states to match expenses to revenues and the attitude of legislatures in balancing the budget)

4. State debt per capita (indicates the future tax burden and the threat of higher future taxes)

5. State spending per capita (indicates the spending tendencies of state governments, which measure the political climate and that climate's impact on the economy)

6. Energy cost per million BTUs (indicates the cost of energy for individual states)

7. Labor union membership (non-agricultural) as a percent of total labor (indicates the degree of unionization in the workforce, and the relative impact of the unionized segment on labor relations, productivity, and legislative and political power)

8. Average weekly manufacturing wage (indicates the relative level of wages and explores the relative cost of labor)

9. Percentage change in average weekly manufacturing wage (indicates the yearly trend in wage rates, which is a measure of the cost of doing business in the state)

10. Person-hours lost per year because of work stoppages per non-agricultural worker (indicates lost productivity and higher manufacturing costs due to labor problems)

Chapter Seven

11. Workmen's compensation insurance rate per $100 of payroll on selected manufacturing occupations (indicates the relative cost of workmen's compensation in the state)

12. Average unemployment benefits paid per covered worker per year (indicates the relative cost of unemployment compensation for the employer)

13. Net worth of state unemployment compensation trust fund per covered worker (indicates the relative strength of the state unemployment compensation fund and the potential for increased unemployment insurance taxes for employers)

14. Maximum weekly payment for temporary/total disability under workmen's compensation insurance assuming recipient earns the average manufacturing wage in the state (indicates the average maximum claim to be paid by the employer for temporary/total disability)

15. Private pollution abatement expenditures as compared to the value of industrial shipments (indicates the cost of environmental control efforts relative to the amount of industry in the state)

16. Private pollution abatement expenditures per capita (expresses the size of the pollution control effort as compared to the size of the population; the higher the cost, the greater its impact on the business climate)

17. Vocational educational expenditures per capita (indicates the level of expenditures designed to maintain and improve the training of the work force)

18. State disbursements for highways per highway mile (indicates relative expenditures for highway maintenance and serves as a measure of the quality of transportation, and thus is one of the variables in quality of life)

Source: Alexander Grant & Company

Tech Transfer Costs

As described in Chapter 3, the technology transfer process encompasses several specific tasks that must be performed before, during and after the transfer of the technology—and each task has its own pricetag. To the extent that the seller or buyer has not prepared in advance specifically for a transfer of the technology, these tasks will piggyback the transfer and increase its cost that much more.

Most inexperienced technology sellers view the transfer process as two or three knowledgeable scientists sitting around a negotiating table to reach an agreement in a gentleman's deal that transfers ownership and makes everyone rich. But the process is much lengthier, and more complex and expensive, than this scenario indicates. Seller transfer costs that must be factored in to the total research and development budget include:

- The extensive amounts of time spent in negotiation: typical negotiations can take from two days to two weeks at an average total cost of $1,000 to $3,000 per day, including hourly rates and travel expenses.

- Legal fees for lengthy consultations on the deal structure and the preparation of agreements range from $5,000 to $25,000 per deal. In contrast, an initial public offering costs $35,000 to $50,000 in legal fees.

- The creation of transfer mechanisms or structures: pure contracts can cost as little as $5,000; complex R&D consortium contracts can cost up to $50,000.

- Travel costs for a transfer involving an overseas participant can range up to $100,000.

- Broker or finder fees: A typical broker fee is approximately 40 percent of revenues received by the seller. Finder's fees range from one to five percent.

Before the seller ever reaches the negotiating table, however, he will have incurred extensive research and development costs:

- Completion of the scientific proof: a laboratory study costs a minimum of $10,000 and can range up to $100,000.
- Completion of the economic proof: these costs also range between $10,000 and $100,000.
- Development of the transfer plan: typically this cost ranges between $5,000 and $20,000.
- Development of promotional materials: fact sheets and other printed materials can cost between $500 and $15,000 for a typical transfer.
- Creation of patent, copyright or trade secret ownership in the technology: the low end is $1,000, up to $20,000 on the high end.
- Accounting/tax advice: between $1,000 and $5,000.

The buyer is not unscathed either. Prior to the transfer, he has incurred the following costs:

- Validation of the data provided by the seller: a complete due diligence costs the same as the economic proof of the seller, or between $10,000 and $100,000.
- Validation of need satisfaction: this cost is included as part of the buyer's due diligence or seller's economic proof expense.
- Accounting/tax advice: $1,000 to $5,000.

After the transfer has been completed and signed, the seller and buyer usually share in some ratio all the costs of the transfer. As a rule, the more complex the transfer mechanism the more numerous the costs. This is true most often when the transfer that occurs through a third entity is established solely for the purpose of facilitating the transfer. The cost of creating a third-party corporate entity, which can be an R&D consortium, can be up to $50,000; in addition, the cost of administering the entity is a percentage of the expense incurred to complete the tasks performed by the consortium. Typically, the management fee can range from two to 15 percent.

Accounting Considerations

It is beyond the scope of this book to describe in detail all of the tax considerations that apply to technology transfer. In most cases, the accounting and tax considerations are complex and variable enough that it is considered prudent to seek outside accounting and tax expertise at the onset of the transfer process.

Topics that should be discussed in depth with an accounting professional include matching income and expenses, the passive nature of royalty income, and qualification for government incentives in the form of tax credits. Another important topic to discuss with a CPA is the timing of revenue and expenses.

Seller Financing

Most transfer pricing strategies do not allow the seller to recapture all initial investment in the technology at the point of transfer, despite the fact that most sellers believe they can walk into the negotiating room with a widget and walk out with a sack of money. The wisest rule of thumb is that the sack of money is almost always smaller than anticipated.

Here is why.

Before the transfer, the seller has incurred certain costs in pushing the technology through the life cycle—which he wants to recapture in the transfer investment. If the seller has self-financed the research and development, he can allow any livable amount of flexibility when the transfer deal requires a long-term payback, for example, from a strapped buyer.

But if the seller has borrowed the startup capital to advance the technology to the transfer table, his deal selection can be restricted to cash on the table, depending on the requirements of his lender. When the deal has to be cash on the table, the seller usually will get a much lower price. His loss can include all or almost all of the product's upside potential; instead the seller will, in effect, be reimbursed for his investment in the technology to date, plus a variable premium.

A hidden cost for the seller, which reduces the amount he walks away with, is the cost incurred for the transfer itself. If a number of staff people are deployed for transfer activities and are not available for income-producing work, normal operating cash flows can be reduced significantly.

A second hidden cost is the potential void left by the development of the technology undergoing transfer. Often, an inexperienced seller will have been working toward transfer for so long a period of time there may be no other projects in the pipeline. In effect, the seller is unemployed and must endure the process of "job-hunting" for a new development project and start all over again.

Buyer Financing

The buyer can experience potentially more difficult financial conditions than the seller in a transfer. Every dollar the buyer pays to the seller at the time of the transfer is one less dollar available to complete the life cycle and get the technology to market. In fact, this hazard is so significant that it is the primary reason why the seller does not walk away from the transfer table with a sizeable chunk of money in most cases.

For one, the buyer may have to finance the license fee paid at the time of transfer, particularly if the deal structure requires a lump-sum payment without future royalties. In effect, the buyer is forced to pay the full opportunity value of the technology as of the date of the transfer.

The transfer becomes even more expensive if he has to finance the tasks remaining in the technology life cycle. Depending on the nature of the tasks, their level of completion, the cumulative risk involved in those tasks, and the amount of capital they require, the buyer may have to finance commercialization through several sources.

Capital and equipment financing needs can be met by standard lending institutions such as commercial banks; capital for operating expenses often is solicited from venture capital firms.

If these sources turn him down, the F&F or friends and family market, and the white-knight market, generally are his last alternatives.

Types of Financing

There are four standard forms of financing for technology transfer:

"State Investments in Technology—1988"

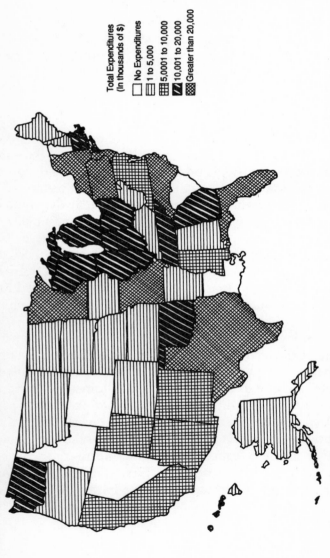

Total Expenditures
(In thousands of $)
- ☐ No Expenditures
- 1 to 5,000
- 5,0001 to 10,000
- 10,001 to 20,000
- Greater than 20,000

Source: Minnesota Office of Science and Technology

Sophisticated technology entrepreneurs often plan to develop and commercialize their products in states that offer generous technology funding in order to increase local tax and revenue bases, and offer higher job-creation opportunities.

Chapter Seven

- Debt: approximately 10 percent of deals
- Equity: approximately 20 percent of deals
- Profit-Sharing: approximately 10 percent of deals
- Royalties: approximately 60 percent of deals

These four options, which have distinctive advantages and disadvantages described below, often are used in combination.

Debt

Technology transfer is not considered a sound basis for outside debt financing by most lenders. Typically, debt financing follows income generating activities, in which the cost of money in the form of interest is paid from income to avoid an ever-increasing debt load.

Lenders do not view transfers as income-producing events because the technology usually is incomplete and the prospect of income lies too far into the future. And to the extent that the economic proofs have not been completed, a realistic assessment of commercialization risk generally is unavailable as well.

Debt financing, however, can be available to a buyer in certain situations. For example, the seller may want to cash in on transfer deals immediately by permitting the buyer to make installment payments.

In effect, the seller has financed the buyer and wants to discount the note to a third party. The seller usually will try to convince a lender to provide bridge financing to cover the costs of a transfer if there is the imminent prospect of cash payment. Typical third parties are venture capitalists; commercial banks generally will not provide bridge financing of this type.

The buyer sometimes can obtain debt financing to cover the capital equipment costs of manufacturing a technology if the equipment is not so unique that it has no resale value. But for the most part, other buyer costs are considered by lenders to be so intangible that debt financing is too risky.

"Funding Sources and Expenditures"

- The federal government funds 48 percent of industry R&D.
- Industry funds approximately 48 percent of its own R&D.
- The Department of Defense/Mission Agency R&D funding has grown about 12 percent annually over the past five years.
- About 28 percent of R&D funding nationwide is devoted to new products, 50 percent to the improvement of existing products, and 13 percent to manufacturing practices.
- Small R&D firms rely more on private-sector sources of R&D capital than on public-sector funding.

Source: The Industrial Technology Institute, 1989

The above graph indicates from what sources large percentages of R&D in the United States have been funded since 1977.

Equity

Equity financing is used more often than lending to fund a transfer for either party because it yields larger returns over a longer period of time to compensate for the high risks involved. An equity investment in a seller is considered good practice because it carries the expectation of ongoing payments from the buyer.

An equity investment in a buyer looks like it will pay off in future returns to the investor when the technology is commercialized. Typical equity investors are most likely to be venture capitalists or strategic corporate "partners."

Sometimes an equity investment is made directly in the technology as part of the royalty stream, although this generally limits the size of the investment. A direct equity investment in the technology is made when the technology looks stronger than other products in the buyer's portfolio or when the buyer wants to participate in potential revenue instead of profit.

This strategy also is used when the buyer wants to avoid the regulations that impact the sale of securities. For example, if the buyer has one "hot" technology and several marginal technologies in his portfolio and the investor does not want to dilute his investment among the marginal technologies, the investor will put up the capital for the acquisition with a royalty override.

Profit-Sharing

Many transfers use a profit-sharing strategy in which risks and profits are shared by the buyer and the seller. Profit-sharing can be a joint venture or the creation of a new entity in the form of a partnership or cooperative. In profit-sharing, the buyer and seller, in effect, obtain financing from the contribution made by the other party.

The buyer usually is the primary beneficiary of this form of financing because payments for the technology are tied to actual sales and limited to net earnings. In this way, the buyer has deferred payment for the technology and has shifted risk, at least partially, to the seller.

Royalties

Royalty financing is very similar to profit-sharing because the buyer has obtained financing for the purchase price by deferring payment until revenue has been generated. Unlike profit-sharing however, the seller takes a cut of all revenue whether or not the venture is profitable. About 60 percent of all deals pay some form of royalties to the buyer as compensation.

"Scope and Performers"

- $100 billion has been spent in applied R&D, $85 billion in development.
- 80% of all development and 50% of all applied research has been conducted by industry.

- The top 100 firms do 79% of industrial R&D, 4 firms do 20%; 89% of all industrial R&D is done by firms with 5,000 employees or more.
- More engineers than scientists are used in technology transfer development: industry employs 80% of the engineers and 48% of the scientists.

Source: The Industrial Technology Institute, 1989

The largest corporations in the United States—resource-rich with capital, personnel, technical expertise, facilities, equipment and industry linkages—account for the largest percentage of all U.S. R&D performed to date.

Sources of Financing

Seller

In most transfers, the seller or transferor bears most of the transfer costs, reflecting the current trend toward seller responsibility and the reluctance of the buyer to use capital to sell himself.

Buyer

A buyer seldom finances a transfer unless the seller is in a weak financial position. That is because he pays most of the sales costs—except buyer identification—to validate the technology in the form of due diligence. If the buyer does finance the transfer, then, it will cost the seller a bundle: he will extract a price differential that is sure to reflect his assumption of risk if the technology fails to perform or commercialization can not be completed.

Venture Capital

The good news is that venture capitalists promise to become even bigger players in the 1990s than they were in the 1980s when "technology capitalists" first specialized with boutique firms. The bad news is that their investment in technology still is governed by the standard investment criterion they have always used: management, management, management.

Even if the seller has a superstar technology—one that will yield 30 percent annual compounded returns in five to seven years and 10 times the initial investment—the transfer will not succeed unless management can get it to market.

New York-based Hambrecht & Quist is a technology-oriented venture capital firm; other firms such as Columbine Venture Fund Ltd. in Englewood, Colorado and Norwest Venture Capital in Minneapolis are traditional venture firms that also finance some technology-oriented portfolio companies.

Banks

The high risks associated with the development of new technology almost preclude banks as a source of financing the costs of technology transfer, with the exception of buyer promissory notes. The only banks that may consider financing all or part of a technology transfer deal tend to be money-center commercial banks such as Manufacturers Hanover and Citicorp in New York City.

Government

Another piece of good news for the 1990s is that many state government agencies are exploring technology investments and transfer financing. Normally, government programs that finance private-sector businesses are structured very strictly according to collateral and cash flow requirements.

But in this decade, many state economic development programs are accepting a higher risk profile in their potential investments and increasingly are willing to finance technology acquisition that attracts more companies and job creation in their jurisdictions.

State agencies such as the Pennsylvania Technical Assistance Program and the Ohio Technology Transfer Organization have financed portions of technology transfer deals. On a federal level, the Federal Laboratory Consortium also has supported technology transfer into the private sector.

Price of Money

The most standard formulae for pricing money in a transfer represent the relationship between risk and reward; but these formulae fail or fall by the wayside as the level of risk or the prospect of reward increases. Transfer pricing generally is conservative, to avoid the astronomical prices that place high revenue demands on the technology and cause it to fail. Standard formulae used to calculate the cost of money in technology transfer deals include the prime rate and the current Internal Revenue Service rate.

For each task in the life cycle of the technology, there is a cost and a reasonable profit margin for its performance. The price of money, then, dilutes that profit margin in a very direct way as a straight deduction. Any transfer price that matches or exceeds a rational profit margin is too high and to be avoided by the experts.

Another factor in the pricing of money is the cost of obtaining the money. Beyond the obvious cost in terms of interest rate, there is a pricetag for seeking, negotiating and receiving the financing. This cost can be very low in a tight, well-run operation when documents that support financing requests are readily available, financing sources have been established previously and origination fees are kept to a minimum.

But if the company is not efficient, these costs can rise to an exorbitant level, including the cost of creating an offering circular, securities and tax opinions, management time to look for money and premiums paid on delivery—which can range from $35,000 to $50,000, or approximately the cost to create an initial public offering.

In fact, these costs can be so high that they exceed the cost of interest. Both the interest rate and the acquisition costs are deducted from profits.

Financing and the Deal Structure

Seller's Perspective

The objective of the seller is to obtain maximum price for the technology, as gross revenue received from the sale of the technology less direct costs and a reasonable profit to the buyer for tasks performed:

Average benefit × number of users − total commercialization costs = projected market price

A cash payment at transfer is advantageous because it removes all risks from the seller. But if the transfer price is set too low, it fails to estimate future sales accurately and the buyer reaps most of the upside potential.

This is the most common fear of sellers who generally believe they sell cheap; but the seller can fail to realize that by underpricing, he has avoided the risk that actual sales may be less than projected. In that case, he has walked away with a reasonable profit and the buyer has bought into a loss.

In lump-sum cash payments, the net present value (NPV) of the technology's potential is reflected as a time/value of money:

> *NPV is based on the presumption that if money is received over a period of time in the future it will be equal to a lesser sum of money today—the difference being interest rate or value placed on the use of the money.*

If the seller wants a royalty or installment payment arrangement, the deal structure should be compared to cash in the bank drawing interest. Sellers often work with a royalty arrangement because they perceive it as the solution to forecasting more accurately the commercial future of the technology.

If the seller maintains a relationship with the buyer after the transfer, it is presumed that accurate sales projections are unnecessary. But without projections, the seller is not considering all the alternatives correctly.

From the seller's standpoint, cash is preferable to a royalty that is based on unreasonably high projections—especially buyer projections—because this indicates the buyer's inability to evaluate the size/dollar value of the market and, potentially, his inability to commercialize the technology and make it profitable due to lack of management skill.

Royalty structures make the seller dependent on the buyer, a vulnerability that can reach down into the buyer's management team and its financial resources: a shortfall in management or resources has a direct adverse impact on sales, and hence, royalties.

The seller also may consider profit-sharing because it allows him the advantage, over a royalty arrangement, of impacting the technology's continued commercialization. To be an effective profit-sharing strategy, the seller should retain the capacity for exerting total control in the event of buyer failure.

If he does not have control, the seller becomes the victim of poor performance on the part of the buyer: sales may increase, but profits are lost or diluted.

So the seller should not lose sight of the need to recapture as much of his investment as possible at the point of transfer. Any arrangement that requires royalties or shared profits puts the investment at risk.

The bottom line is the seller may be better off with a minimum payment that ensures recapture of the initial investment and a reasonable profit—without precluding other options.

The Buyer's Perspective

Most buyers want some form of financing unless the seller needs a fire sale and the price bears little resemblance to prospective opportunity or costs. As a rule, this is because the value of the technology far exceeds the amount of cash a buyer can pay upfront. The buyer, therefore, usually insists on a price structure that allows deferred payments, particularly when they are tied to future sales:

> 5 to 10 percent of projected royalties are typically paid upfront in an initial lump-sum payment and the balance is paid to the buyer upon actual sales.

He also wants to avoid transfer structures that create fixed liabilities in the future, which he may not be able to meet. For example, the buyer promises to pay a fixed number of dollars within a fixed period of time, say $50,000 per month for three years.

This is especially onerous to the buyer when an installment payment program to the seller must continue even after the technology has failed. Another example is a royalty program that puts the buyer at a disadvantage to a competitive technology which does not carry a royalty burden—creating a sales price disadvantage.

An inexperienced buyer may promise to pay a royalty to the seller by raising the consumer price to absorb the cost of the royalty, or he will add the cost of royalties to the cost of commercialization.

The buyer's bottom line is the recognition that no matter which transfer structure is used, he has additional and substantial costs; and every dollar that he invests is subject to the risk of failure even after the technology reaches the marketplace.

Forms of Compensation

There are three forms of licensing compensation: (1) licensing fees, including royalties (rate × basis), and margins on supplied parts and products purchased from the licensee; (2) one-time fees, including front end or lump-sum payments and service fees for technical assistance; and (3) returns in the form of grant-backs and licensee equity.

The most common form of compensation is a royalty fee based on some level of licensing activity, usually the turnover produced by the licensee because of its stability over time and its ability to deliver returns even when the licensee is not profitable.

Profits also can be used as a basis for calculating royalties, but they fluctuate widely and can be manipulated too easily in the accounting department. And if profits are used to determine the royalty rate, the licensor receives no royalties if the licensee does not sell product.

The amount of actual royalties or the royalty rate is contingent, for the most part, on the nature of the technology licensed and on the negotiating abilities of the buyer and seller, or licensee and licensor.

Most royalty rates fall between 1 and 5 percent of turnover; occasionally they are divided into minimum royalties and running royalties. Minimum royalties guarantee a minimum income for the licensor on an annual basis. Running royalties provide top-up income that fluctuate with the ability of the licensee to sell product.

Margin also provides compensation on the basis of licensing activity, including the licensor margin on components the seller supplies to licensee and the margin on products the seller receives from the buyer/licensee. This margin also can be covered by a buy-back clause, which allows the seller/licensor to buy back product from the licensee at a discounted price.

Most licensing agreements provide the payment of one-time front end or lump-sum fees which are paid upfront to the seller/licensor to decrease the licensor's risk in disclosing technical know-how, to cover political risk in foreign markets, and to cover startup and training costs incurred in administering the license.

Technical or servicing fees usually are negotiated by the seller/licensor upfront to cover the on-going costs of training and consultation with the licensee. Grant-back clauses stipulate that any improvements made by the licensee to the licensed technology revert to the licensor.

Some licensors want to insure long-term profits by purchasing equity in the buyer/licensee, either partial equity or eventual full ownership of the licensee.

8
Legal Skills

A timely $10,000, spent effectively, covers the cost of establishing ownership in most technologies.

The transfer of technology has been described in the first seven chapters of this book as a transfer of responsibilities for completing the commercialization of a technology; in this chapter, the process is analyzed as a transfer of ownership rights to the technology.

These rights are the legal rights of ownership which give a rightsholder the ability to control the technology by limiting or excluding others from participation in any phase of its life cycle during commercialization. These rights often are referred to as "intellectual property," or assets created by the mind.

Of equal concern is the business structure that owns the technology, which can be a sole proprietorship, a corporation or a partnership. Each of these business structures has certain advantages and disadvantages relative to technology transfer, and some new structures are created solely for the purpose of facilitating a transfer.

Intellectual Property

Intellectual property encompasses the rights inherent in a patent, copyright and trade secret. Each of these forms of intellectual property legally permits the rightsholder of the technology to use a different set of rules to control the technology.

The ability to receive any of these rights is contingent on the characteristics of the technology; but as described earlier, the characteristics of the technology go through rapid change as the technology passes through each life cycle stage. So this means that the form of intellectual property, or choice of legal protection, varies over time according to the stage of the technology life cycle.

Patent

This form of intellectual property protection gives the holder the right to exclusive control of the technology for a period of 17 years. A patent is granted to protect machines, processes, material compounds and designs which are novel and non-obvious. Patent rights are created by registration with and the approval of the U.S. Patent office.

Copyright

This form of intellectual property protection gives the holder the right to control the technology for the life of the author or 50 years, whichever is greater up to 100 years. It does not prevent a competitor from creating a similar work; but it does protect the holder against a competitor who wants to copy his technology directly.

A copyright is limited to works of authorship, which can include literary works, art, movies and sound recordings; it is vital to know that copyright protection does not extend to intellectual concepts. Copyright protection is created automatically when the work of authorship is completed; additional copyright protection is created when the work is registered with the U.S. Copyright Office.

Trade Secret

This form of intellectual property protection gives the holder the right to control a technology secret as long as the information about the technology remains a secret. Trade secret protection does not, however, prevent a competitor from independently creating a similar secret; nor does it prevent a competitor from reverse-engineering the secret from a product in which the secret is utilized.

Trade secret protection is limited to information that is held in secrecy in order to provide a competitive advantage to the holder. Trade secret rights are created by statute, by contract or by business relationship.

Chapter Eight

Licensing

Licensing is the transfer of less than all rights in a technology, and it is this aspect of technology transfer that offers one of the greatest opportunities for creating technology wealth.

The opportunity exists because licensing allows a rightsholder to sell the technology, or portions of it, more than once for one or more applications. In the same way that a manufacturer can create a large number of identical products and make a profit on each product unit, the technology transferor can profit by creating a number of licenses.

The trick is to understand how licensing rights in the technology can be separated and sold separately. Even after separation, additional sales can be created by limiting the rights to specific markets.

Rights of Use

The holder of a patent, copyright or trade secret holds certain privileges or rights, including:

- the right to use the technology
- the right to transfer the technology
- the right to replicate the technology
- the right to modify the technology

Right of use is exercised by the consumer or end-user at the end of the technology life cycle when the technology has been commercialized. Right of transfer allows a technology transfer to occur; the technology holder freely can transfer all or part of the remaining rights to someone else.

Right of replication is exercised by the manufacturer, which produces multiple copies of the technology. Right of modification allows the technology to be developed further through research and development, or through integrating the technology with other products or services.

Splitting Rights

A simple split of rights occurs when one of the primary rights—the right to use, transfer, replicate or modify the technology—is licensed.

To some extent, this form of licensing adapts to the life cycle of the technology, allowing for the manufacture, distribution, sale and use of a technology product.

A more sophisticated split segregates different applications of the technology by allowing the technology to be used to address different problems in separate markets serviced by different industries.

Limiting Rights

With or without actually splitting the technology rights, restrictions may be placed on the grant of rights which appear to be another form of splitting. The grant may be limited to a period of time, a geographic area or a market niche; in fact, the variety of limitations can be virtually infinite. Limitations typically are used to define a specific market segment.

Nondisclosure Agreements

The wisdom of using nondisclosure agreements has been established almost without argument in the technology—and many other—industries. Entrepreneurs who plan to obtain copyright, patent, trade secret, license or trademark protection for any aspect of the technology under development, from the very beginning at concept stage through the use stage, should hire legal counsel to develop specific nondisclosure agreements relative to the technology for everyone from employees to outside investors.

Although many entrepreneurs are familiar with the reasoning behind the use of a nondisclosure agreement, many more neglect to use the agreement at the appropriate times, either through honest omission, fear of offending a potential investor or ignorance of such marketplace practices as reverse engineering. It is imperative that a technology seller use nondisclosure agreements to protect his rights of ownership when disclosing aspects of the technology to potential business partners, scientists, employees, investors, media representatives and outside parties who are interested in the development of the technology.

Without the documentation provided by signed nondisclosure agreements, the technology seller has little recourse when an outside entity elects to further develop, sell or use the technology without the seller's agreement.

See Appendix C for a sample nondisclosure agreement which can be used as a model for technology entrepreneurs.

Contracts

Most technology transfer mechanisms are documented by a written contract which establishes the terms and conditions for the transfer and operates as a custom set of rules for the conduct of the signers while the transfer is taking place.

Agreement

For a contract to exist, there must be an agreement between transfer participants. The buyer and seller must have a mutual understanding about the transfer which is best documented in writing. In fact, the written contract is so important that many forms of intellectual property protection do not recognize a transfer of rights without a written document to indicate the change. In addition, a well-written document also helps the buyer and seller demonstrate their understanding of the agreement.

Performance

Within every contract, the buyer and seller each are bound to a set of legal obligations and a corresponding or reciprocal set of legal rights. In order to successfully complete the contract, both parties must perform their obligations fully. Failure to perform by either party results in a breach by default of the contract and can be the impetus for future litigation.

Fraud and Misrepresentation

Care must be taken during every technology transfer to avoid misstatements of fact because they can lead to claims of fraud or misrepresentation. These claims can subject the speaker to damages arising from the failure of the contract.

A clear understanding of fraud and misrepresentation is crucial because fraud and misrepresentation are not limited to actual oral statements; they also include failure to provide information when its absence can be construed as misleading.

Fraud and misrepresentation are of particular concern also when the seller's technology is incomplete and/or its characteristics are not fixed, or when many of its elements are unknown and/or without capability of proof.

Incomplete technology can lead to continuing communication problems about the potential inherent in the technology when the oral statement about its capability is taken as fact or as a guarantee.

Business Structures and Relationships

Technology transfer participants are subject to the laws of the geographic territory within which the transfer takes place, which can include the laws of one or more states in the United States. These laws govern the conduct of both transfer parties. Special forms of relationship invoke specific laws for each form of relationship.

Employer/Employee

In almost all situations, technology that is created by an employee during the course of his employment by another entity is owned by the employer. However, this general rule can be altered by contracts with third parties, by the conduct of the employer or by an employment agreement.

Independent Contractors

Most technology transfers occur within the framework of an independent contractor relationship, which is governed by distinct independent contractor rules of conduct within the transfer agreement. This agreement can be a license or assignment.

Partners

Some technology transfer mechanisms create a partnership, which can be temporary and of limited scope like a joint venture, or can be extended over time like a research and development cooperative. The unique obligations and rights of partners are delineated in a special partnership agreement.

Government Regulations

Technology transfer activities are subject to control by local, state and federal governments for the health, safety and welfare of U.S. citizens. So it is important to know that any of these laws and regulations determine and/or restrict to a great extent how the technology is transferred.

International

(See Chapter 9 for a full description of international legal issues.) It is imperative that technology entrepreneurs stay abreast of fast-breaking changes in legal practice around the world. Following is a partial sample of worldwide legal cases that were covered in 1988.

"Offshore Legal Issues Reported in 1988"

- Leader Radio Co. Ltd. v. Wing Hing Wires & Cable Factory Ltd. (Hong Kong): The Supreme Court refused to accept a private investigator's evidence of trap orders in a passing off case.
- The Gillette and Yardley cases in India, in which the famous foreign plaintiffs succeeded in actions for passing off even though they did not have a business in the country.
- The Rhone-Poulenc case in Malaysia: the first fully litigated case under the new Malaysian Patent Act.
- The Winthrop/Sterling case in Malaysia: the plaintiffs failed in an attempt to use the trademark law to prevent parallel imports of Panadol.
- The Davidoff case (Singapore): the court took a generous view of group reputation and prior use.

- Kawada v. Suwaki Inc.: the Japanese Supreme Court ruled that a trademark for noodles was infringed by the use of similar characters to identify a restaurant chain.
- Lego v. Tyco: decided by the UK Privy Council on appeal from Hong Kong, clarifying the definition of "design" and the scope of copyright protection in derivative drawings.
- Patent disputes litigated by Bristol Myers, Squibb and Bayer in Korea: the American companies resorted to s. 3-1 of the U.S. Trade Act. The West German company (Bayer) was able to successfully oppose the grant of a process patent using only Korean domestic law.
- The Parker case (Korea): the Korean court gave wide protection to a famous foreign mark.
- The Harris and Ramada Inn cases in Thailand, where the famous foreign plaintiff was less successful, and the Christian Dior case, in which the famous foreign mark was protected.

Source: IP ASIA, May 1989

IP ASIA, the Hong Kong-based legal report published 10 times annually on intellectual property, marketing and communications law as it is practiced in Asia, has highlighted a number of significant legal cases involving American companies doing business overseas which should serve as a guide for potential technology transferors.

This unique publication is important because it provides legal alerts and global updates that signal important trend changes in the practice of intellectual property law, changes about which both transferors and their legal counsel should become and remain aware. It also apprises U.S. citizens abroad of the U.S. government's policies regarding such important issues as antitrust issues in the global marketplace.

Antitrust

One of the federal government's primary concerns is the continual promotion of free enterprise and capitalism. To this end, many U.S. laws and regulations attempt to control restrictions of trade on the part of its own citizens and foreign traders who trade both within the United States and outside its borders.

These public trade policies—which include tariffs and quotas, among other trading strategies—often conflict very directly with the commercialization of technology and with intellectual property rights.

This conflict has become so direct that it is recognized as a major stumbling block to the successful completion of technology commercialization, which requires at least some degree of exclusivity without interference or competition from the government. The R&D laboratory can become a battleground for the federal government when a technology must be developed on an exclusive basis.

In 1984, Congress passed the Federal Competitiveness Act, which allowed competitors to join together to participate in common research efforts without violating current antitrust laws. But some antitrust restrictions remain, including "price-fixing" and "product tying."

Price-fixing is the attempt to impose improper price controls; i.e., when a manufacturer or distributor dictates the price at which the seller must sell the product to the consumer. Suggested prices are permitted as long as the dealer cannot be punished for using alternative prices.

Product tying is the forced purchase of another product; i.e., the desired product is sold only on the purchase of another, often unneeded or undesirable, product.

Dispute Resolution

Not all technology transfers are successful; the failures are due often to known risks over which neither the buyer nor the seller had control. Other transfers fail when one or more of the participants in the transfer neglect to fulfill their contractual obligations.

Many transfer failures result in a dispute and a claim for damages by either the buyer or seller, no matter why the transfer failed. So several mechanisms have evolved over time and usage by which to resolve transfer disputes.

Negotiation

This form of dispute resolution is limited to the participation of the parties involved in the dispute, without the involvement of anyone else. Negotiation allows the disputing parties to retain complete control over the dispute and to take any action to which both parties mutually agree.

Mediation

This form of dispute resolution permits the participation of an outside third party as a mediator. The mediator is selected by the mutual agreement of the disputing parties or by the rules established by the terms in the transfer agreement.

Mediators typically are chosen for an outside, objective viewpoint, technical knowledge and experience with the technology and/or industry. Mediators usually provide recommendations only, which the parties are not required to accept.

Arbitration

This form of dispute resolution requires the participation of an outside third party as an arbitrator, who presides over the dispute in the role of judge and jury. The arbitrator resolves all issues of fact and law in the dispute, and the award or decision of the arbitrator is binding on both of the disputing parties.

Litigation

This form of dispute resolution requires that the dispute be settled in court using the governmental judicial system. In disputes, it is the only appropriate forum in which to resolve issues of ownership of intellectual property rights.

Examples of international dispute resolution bodies are:

>The International Chamber of Commerce
>38 Cours Albert ler.
>75008 Paris, France
>Telephone: 261-85-97
>
>American Arbitration Association
>140 West 51st Street
>New York, NY 10020
>(212) 484-4000.

9
International Transfer

Technology transferors can expect to comprise more than 10 percent of the total global market in the 1990s, which was valued at $31 trillion in 1988.

In the 1990s, it will become increasingly difficult to complete a technology transfer without taking into consideration the international dimensions of the transfer, whether or not the technology buyer or seller is based in the United States and whether or not the end-users are U.S. citizens.

Because of the velocity at which trade is occurring around the world and because national borders are increasingly blurred in the financial, trade and investment sectors, technologies that are developed, bought or sold domestically can appear in the global marketplace almost instantaneously.

So all technology either is sold or should be considered a potential major export for the United States—as well as for other nations—due in part to U.S. leadership in reasearch and development, in part to the fact that Third-World nations can use technology that is disallowed within the United States or is obsolescent here, and in part to the capability of other nations to perform more economically the manufacturing stage of the technology life cycle.

To be considered profitable and effective, a license transfer, for example, should be viewed as part of a global planning strategy that has been assimilated to exporting and foreign production, where appropriate for small business owners that hold technology rights.

Licensing agreements can be a complement to a foreign subsidiary in a country that allows royalties to be paid to the parent company in lieu of dividends. In this case, the license actually can protect business that is created overseas. In fact, a major portion of global trade results from global licensing agreements that function in this way.

According to a 1986 OECD report, the total number of licensing agreements among the largest industrialized nations increased in constant terms by 2 percent annually between 1975 and 1983.

Those payments (measured in 1980 prices and exchange rates) jumped form $9.7 billion in 1975 to $11.4 billion in 1983. The sale of products manufactured under independent license was equivalent to between 5 and 10 percent of sales from foreign production, the majority of which came from the United States, the United Kingdom, France, Germany and Japan.

Page 157 shows a comparison of a sampling of Japanese and U.S. companies that spend high percentages of annual revenue on research and development activities in pursuit of lucrative license agreements in the global marketplace.

International Transfer Protection

Several international conventions and treaties have encouraged the worldwide integration of patent protection law in order to decrease the exponential growth rate in infringement and other legal problems that resulted from increases in international trade and transfer in the past century.

The first was the 1883 Paris Convention for the Protection of Industrial Property, which was signed by approximately 85 nations to ensure uniform protection of industrial property worldwide. Others are the 1977 European Patent Covention and the 1978 Patent Cooperation Treaty.

Although the first international treaties addressed only protection for patent rights, several other forms of intellectual or industrial property rights have become significant players in world's technology transfer marketplace.

Global Licensing

Like domestic licensing, the global license of a technology owner allows others to develop, use or sell the technology under certain

"Comparison of Japanese and U.S. Technology Firms By R&D Rates"

Japan	Sales	Profit	Income (%sales)	R&D	R&D (%sales)	R&D Growth	Captial Expend.
Toshiba	$28.6 B	$468 M	1.7%	$1.74 B	6.08%	7.4%	NA
Mitsubishi Electric	$18.9 B	$178 M	0.94%	$0.82 B	4.33%	5.9%	$1.75 B
Sumitomo Electric	$4.4 B	$100 M	2.27%	$0.18 B	4.05%	4.84%	$1.01 B
Hitachi	$23.4 B	$521 M	2.23%	$2.10 B	8.97%	−1.4%	$6.76 B
Matsushita	$38.6 B	$1.3 B	3.37%	$2.17 B	5.62%	5.07%	$1.4 B
NTT	$45.3 B	$1.9 B	0.43%	$1.45 B	3.21%	5.18%	$5.42
Fujitsu	$13.7 B	$256 M	1.87%	$1.47 B	10.76%	8.53%	$2.76 B
Furukawa	$4.0 B	$52 M	1.3%	$0.09 B	2.3%	7.89%	$0.61 B
NEC	$21.9 B	$204 M	0.93%	$3.46 B	16.0%	7.0%	$2.0 B
United States							
IBM	$59.6 B	$5.8 B	9.7%	$4.4 B	9.89%	41.9%	$23.4 B
General Electric	$49.4 B	$3.3 B	6.9%	$1.2 B	2.42%	15.6%	$2.3 B
Du Pont	$32.5 B	$2.1 B	6.7%	$1.3 B	3.69%	16.2%	$4.2 B
United Technologies	$180.0 B	$659 M	3.6%	$932 M	5.17%	0.2%	$875 M
Xerox	$16.4 B	$388 M	2.4%	$794 M	4.84%	43.1%	$1.4 B
Westinghouse	$12.5 B	$822 M	6.6%	$706 M	5.64%	-17.1%	$421 M
Rockwell Int'l.	$11.9 B	$811 M	6.8%	$1.6 B	13.44%	42.5%	$555 M
Allied Signal	$11.9 B	$463 M	3.9%	$647 M	5.43%	66.3%	$602 M

Source: JTech and Mary Rouvelas for "New Technology Week."

One reason licensing and royalty growth rates are picking up is the increase in R&D expenditures indicated for some of the world's top technology firms. Spending significantly greater proportions of their revenue on research, these firms must go to the global marketplace in order to find profitable outlets for their technology development. Although the Japanese firms in the chart above would not fare well on Wall Street, they ignore quarter-by-quarter earnings expectations to spend considerable sums for long-term growth and profit.

restrictions for a certain period of time in exchange for some form of agreed on compensation. In international licensing, there are three distinct types of technology licensed.

(1) Process technology is the systematic knowledge utilized to operate equipment and machinery to produce certain products. (2) Product technology is the systematic knowledge utilized to produce certain products. (3) Management technology is the managerial knowledge utilized to enable a company to operate its business in a way that achieves its goals.

Rightsholders or companies that license as suppliers of technology—the licensors—or recipients of technology, the licensees, are involved either in "outward licensing" or in "inward licensing."

Global Patenting

A global patent is a temporary monopoly grant by a government to the technology owner or developer of a novel technology or product. It has an expiration date, which in the United States is 17 years and in Europe, for example, is 20 years. A patent owner has the exclusive right to use an invention and to sue an infringing competitor of this right.

In order for the technology owner to acquire adequate protection in the global marketplace, a patent application must be filed in every country where a patent may be utilized, including the country of origin of the seller, the buyer, the distributor and the end-users.

In fact, a technology owner cannot enter into a licensing agreement with a foreign company or nation unless the patent is registered in that foreign country.

International Trade Secrets

An alternative to a license or patent is the trade secret, which can consist of a drawing, paper, design or formula containing confidential information about a technology. A good example of an international trade secret is the composition of Chartreuse liqueur, which has been the confidential possession of the Carthusian monks for more than 400 years.

The owner of trade secret rights does not have the exclusive right to exploit his technology, nor does he have the right to sue anyone who independently develops the technology.

Trade secret protection applies only as long as the secret remains secret. If a competitor develops the same secret, he also has the right to use the technology. It is important to note that most nations have no specific statutory or other laws to protect trade secrets in the absence of a contract.

But there are two ways to protect trade secrets: (1) legally contract for confidentiality between the technology owner and other users, in the form of a licensing agreement for external users, or (2) legally contract with internal users in the form of an employment contract. Both of these contracts then permit lawsuits for damages under common law.

Despite this disadvantage inherent in trade secrets, they do present one large advantage to the technology holder: in contrast to patents, trade secrets extend for an indefinite period of time and can generate royalties from licensing well beyond the 17- and 20-year period of patents.

In this way, technology owners not only can generate nearly unlimited income, they also can protect their information indefinitely from competitors—in contrast to the disclosure required to register a patent, which often results in competitive engineering around the patent.

Know-how is a form of trade secret if the information is not available to the public at large. Know-how is distinguished from a trade secret by patentability: a trade secret can be patented if the technology holder so desires; know-how normally is unpatentable.

As a form of trade secret, know-how must be protected in the same way—by legal contract between the technology holder and the end-users, which can be protected by common law.

International Trademarks

Trademarks also are important in the international marketplace as visual or aural marks, devices, words, symbols or combinations that help the owner and end-users distinguish among a global array of competing technologies or products.

Needless to say, trademarks are significant arbiters of quality, and that quality control of the licensor over licensees is an important capability to retain via trademarks.

Trademarks carry property protection rights similar to patents under the Paris Covention and the Madrid Agreement through a registration with the Geneva-based World Intellectual Property Organization. These marks also are protected in many nations by unfair trade laws that protect trademark holders from unfair competitors who "pass off" their products as those of a different company.

Trademarks often are included as part of patent, trade secret or know-how agreements in order to prolong the life of the technology by piggybacking it to a trademark, which is protected for an unlimited period of time. The limited intellectual right does not run out as long as the technology produced under the patent, for example, is identified in the minds of end-users through the use of the trademark.

International Copyright

The importance of the copyright in the global marketplace cannot be overemphasized. It gives the technology holder protection in everything from marketing and promotional literature—including advertising and public relations materials—to technical documents and drawings, computer software and designs.

The characteristic that distinguishes the copyright from a patent or trademark is automatic protection. Copyright protection exists automatically on the completion of the work; patents and trademarks must be registered before protection exists.

Most copyrights anywhere in the world exist for 50 years after the death of the inventor or originator, with some exceptions, under the protection of the 1886 Berne Convention.

Export Restrictions

The United States has created—and enforces—a rigid set of laws and regulations regarding the transfer of technology to other countries. These laws exist primarily to guarantee national security and to prevent political enemies of the United States from developing conventional or nuclear weapons.

As a result, exporting technology from the United States is subject to the approval of the U.S. Department of Commerce. In addition, weapons and nuclear energy technology are subject to further approval by the U.S. Department of State, the U.S. Department of Defense and the Nuclear Regulatory Commission.

Any one or all of these regulatory bodies can limit severely or forbid entirely the transfer of technology into other nations, usually those considered to be communist or enemy nations. The restrictions that apply to the exporting and offshore technology transfers of finished products also apply to technical data. This impacts the transfer of incomplete technology, which can be anything from products and data to services that utilize products and data.

Another significant aspect of technology transfer is the requirement for a U.S. export license. Most exports require a general license, which is no more than the completion of appropriate forms and requires no pre-approval prior to shipping.

In contrast, more sensitive technology requires a validated license, a permit which must be approved in advance by the Department of Commerce and other agencies, when appropriate. The most sensitive technology as determined by the federal government cannot be shipped overseas except to U.S. allies.

Currency/Foreign Exchange Exposure

The sale of a technology outside the United States does not result automatically in payment in the form of U.S. dollars. A transfer often is completed in the local currency of the foreign buyer or seller, which is subject to on-going fluctuations in exchange rates.

Some countries, such as the Soviet Union and China, either have no exchange rate or have severe restrictions on payment in hard currencies like the U.S. dollar. This requires payment in other forms, possibly commodity goods, in an exchange of goods referred to as "countertrade."

Because international tariff barriers have been almost completely eliminated, trade in the 1990s increases briskly on an annual basis; as a result, so do capital flows, and speculative and financial capital.

To get a perspective on the size of the global market, consider that between 1983 and 1988 the volume of foreign currency trades per business day in the United States jumped from $49 to $129 billion, according to a Federal Reserve Bank of New York survey. On an annual basis, the numbers vaulted from $12 trillion in 1983 to over $31 trillion in 1988.

Because of the continuation of floating exchange rates among many foreign nations, the exposure from foreign exchange risk should be of increasing importance to exporters and technology transferors.

Direct foreign exchange risk usually occurs on the financing or direct sale of a technology, in this instance. If there is multiple licensing, each individual license transaction must be planned and negotiated separately in order to avoid foreign exchange exposure.

There are a number of strategies by which the U.S. technology holder can minimize foreign exchange risk in his transactions with foreign nations through the use of international banks. The foreign exchange arena has become an electronic communications market, 85 percent of which is comprised of the major international banks. Foreign exchange brokers are intermediaries who establish market prices.

Although in 1988 over 80 percent of the total foreign exchange trading took place among international banks, the market would not exist without corporate commercial business—from multinationals, exporters and transferors of technology. This demand for foreign exchange activity on the part of corporations is the foundation of bank-trader profits and losses, through such transactions as:

- Hedging
- Foreign Exchange Options
- Spot Contracts
- Swaps
- Outright Forwards

Cultures and Customs

To be successful, an international transfer must take into consideration the cultures of all participants. That usually means that both buyer and seller—and particularly the American participant—must be prepared to compromise his customary form of soliciting, negotiating

Chapter Nine

and contracting deals in order to adopt practices that are acceptable to all parties and complete the deal.

Many transfer negotiators are surprised to discover that misinterpreting cultural differences, or the inability to traverse them, can kill a transfer as quickly as the inability to negotiate agreeable terms.

One common variance from American business practice is the lack of dependence on written contracts in many parts of the world. In countries that de-emphasize the written word, contracts are used only as a tool for expediting the transaction because they place greater importance on relationship than on the written word.

In contrast in the United States, great reliance is placed on the contract without regard to the individuals behind it. This reliance is supported by a legal system unparalleled in the world. The emphasis is placed, incorrectly, on enforcement instead of on simple performance. Outside the United States, reliance and trust must be placed on the business partner.

These countries, therefore, have to depend heavily on their legal systems to help enforce the terms of the contract, an undesirable result which can become very costly to both parties.

Take, for example, the vast corporate cultural differences between Japan and the United States. Japan is an important potential buyer in American technology transfers, and may continue to be important in the 1990s. One reason is its very significant economic growth since World War II: Japan's growth rate exceeds the global average—in 1987, Japan's GNP was 15 percent of the global GNP, up more than 5 percent from 1977—and big increases in yen appreciation have pumped Japan's net external assets to more than $270 billion, which is the highest in the world.

This kind of growth has been fueled by several, almost national, management policies that allowed Japan to catch up with America in the post-war era: a no-risk policy that produced successful short-term results; a seniority-based personnel policy that kept most employees from taking excessive risks on the job; a policy of restricted job responsibility and simultaneous personal accountability; a strong orientation toward group decision-making; a policy of homogenizing the corporate workforce so employees with similar traits and abilities are working in tandem toward the same goals; and a policy that favored hierarchical organizational structures to enforce class rigidity.

Since 1985, however, the strong yen has increased the export prices of Japanese products, resulting in the reduced competitiveness of these products in offshore markets, notably in the United States. Japan's exports in 1985 indicated a significant decline in balance, which continued into 1986, and was the primary foundation of the world trade imbalance at that time.

This situation marked a turning point in Japan's economic structure and in the managment policies that had been used so successfully in the past; two factors that promise to complicate the negotiation of a technology transfer with a buyer in that country.

Japanese negotiators in the near future probably will present a blend of old and new strategies that include vague decision-making—which worked so well for the Japanese in the past—coupled with speedier American tactics that will bring Japan into the information age; more emphasis on software-related departments, as Japan moves to diversify corporate business; and a pendulum effect in importing as Japan attempts to strengthen domestic business by decreasing its dependence on exports.

As the need to diversify its business base increases with economic maturity, Japan will move toward heavy investment in new production facilities. It also will require technically-proficient personnel who can accommodate all the cutting-edge technologies that have been imported into the country in recent years.

This fact will change the emphasis in technology transfer negotiations considerably for American sellers, and should be backgrounded thoroughly by U.S. business owners who anticipate transferring to Japan in the near future.

Also, look for the Japanese to become much sharper negotiators of foreign exchange risk and exposure. Another important cultural difference between Japan and America is evident in Japan's former position on foreign exchange: it was speculative and, in fact, was legally prohibited until recently.

While that attitude certainly will not disappear overnight despite the fact that foreign exchange prohibitions have been lifted, Japan no longer considers foreign exchange speculation as a necessary evil. At best, the Japanese will be ambivalent foreign exchange negotiators in the near-term.

Appendix A

RESOURCES FOR TECHNOLOGY TRANSFER

Associations and Organizations

Technology Transfer Society

611 North Capitol Avenue
Indianapolis, Indiana 46204
(317) 262-5022

This national organization was formed in 1975 to serve as a catalyst to help accelerate the movement of technology out of the laboratory and into the private sector. The group uses a newsletter, a journal, meetings of local chapters, and an annual conference in order to exchange information on the methodology, utilization, assessment and forecasting of technology transfer. Membership is open to technology-transfer professionals, technologists, scientists, engineers, economists, attorneys, venture capitalists, bankers, business people, information specialists, consumer groups, teachers and students interested in the society's goals.

Licensing Executives Society (LES)

71 East Avenue
Norwalk, CT 06851-4903
(203) 852-7168

This international organization promotes technology licensing through publication of *LES Nouvelles*; regional, national and international conferences and meetings; training; and service as an information source.

National Academy of Sciences

2101 Constitution Avenue, N.W.
Washington, D.C. 20418
(202) 334-3486

Association of University Technology Managers

C/O Spencer Braylock
315 Beardshear
Ames, Iowa
(515) 294-4740

This organization is comprised of individuals responsible for technology transfer from universities, hospitals and nonprofits.

Federal Laboratory Consortium for Technology Transfer (FLC)

FLC Administrator
1945 N. Fine, Suite 109
Fresno, California 93727
(209) 251-3830

This federal consortium was formed in 1974 to promote technology transfer by assisting federal laboratories in the development of infrastructure, by facilitating links to the public and private sectors, and by promoting collaboration at the regional and national level. The FLC publishes a newletter and holds training programs for members.

Database and Directories

Technology Catalysts, Inc.

6073 Arlington Blvd.
Falls Church, VA 22044
(703) 237-9600

Technology Search International, Inc.

500 East Higgins Road
Elk Grove Village, Illinois 60007
(312) 593-2111

TECHNOTEC

Control Data Corporation
7600 France Avenue South
Ediman, MN 55435
(612) 893-4640

Appendix A

CorpTech

    Corporate Technology Information Services, Inc.
    12 Alfred Street, Suite 200
    Woburn, Massachusetts 01801-9998
    (617) 932-3939

TechBase

    Technical Insights, Inc.
    32 North Dean Street
    Englewood, New Jersey 07631
    (201) 568-4744

Tech Notes

    National Technical Information Service
    5285 Port Royal Road
    Springfield, Virginia 22161
    (703) 487-4600

Conferences and Fairs

Tech Ex

    Dvorkovitz & Associates
    P.O. Box 1748
    Ormond Beach, Florida 32075
    (904) 677-7033

Technology Transfer Conferences

    Alladin Industries
    P.O. Box 100225
    Nashville, Tennessee 37210
    (615) 748-3108

Invention Convention

    ICS Corporation
    6753 Hollywood Blvd., −212
    Hollywood, California 90028
    (213) 460-4408

Newsletters, Magazines and Periodicals

"Research & Development," Cahners Publishing Company, 275 Washington Street, Newton, Massachusetts 02158-1630, (303) 388-4511.

"Technology Access Report," Technology Access, P.O. Box 778, Inverness, CA 94937, 1-800-733-1516.

"Leading Edge Technologies," IC2 Institute, University of Texas at Austin, 2815 San Gabriel, Austin, Texas 78705, (512) 478- 081.

"Technology Alliances for Competitiveness," International High Technology Group, KPMG Peat Marwick, 303 East Wacker Drive, Box 55, Chicago, Illinois 60601, (312)938-1000.

"New Technology Week," King Communications Group, Inc., 627 National Press Building, Washington, D.C. 20045, (202) 638-4260.

"Advanced Manufacturing Technology," 32 North Dean Street, Englewood, NJ 07631, (201) 568-4744.

"Technology Access," University R&D Opportunities, Inc., P.O. Box 778, Inverness, CA 94937, 1-800-882-7227.

"Technology Review," 201 Vassar Street, Cambridge, MA 02139, (617) 253-8250.

"IP ASIA," Shomei Ltd., 2/F Yu Wing Building, 64-66 Wellington Street, Central, Hong Kong, 5-8458183.

The Journal of Technology Transfer, Technology Transfer Society 611 North Capitol Avenue, Indianapolis, IN 46204.

Books

"From Source to Use: Bringing University Technology to the Marketplace," Alva L. Frye, Editor, American Management Association.

Appendix A

"Guide To University-Industry Research Agreements," Society of Research Administrators, 1505 4th Street, Suite 203, Santa Monica, California 90401.

"Intellectual Property Management," Philip Sperber, Clark Boardman Company Ltd., New York, New York.

"International Technology Licensing," Farok J. Contractor, Lexington Books, Lexington, Massachusetts.

"International Technology Transfer," Howard V. Perlmutter and Tagi Sagafi-nejad, Pergamon Press, Inc., Maxwell House, Fairview Park, Elmsford, New York 10523.

"Lesko's New Tech Sourcebook," Matthew Lesko, Harper & Row Publishers, Inc., 10 East 53rd Street, New York, New York 10022.

"Marketing High Technology," William H. Davidow, The Free Press, 866 Third Avenue, New York, NY 10022.

"Technology Transfer: General and Theoretical Studies, Bibliography of Papers from Feb. 1985 to May 1988," U.S. Department of Commerce, National Technical Information Service,Springfield, VA 22161, (703) 487-4650.

"The Entrepreneur's Guide To Capital" (second edition), Jennifer Lindsey, Probus Publishing Company, 118 North Clinton Street,Chicago, IL 60606, 1-800-426-1520.

"Finding and Licensing New Products & Technology (U.S.A.)"Technology Search International, Inc., 500 East Higgins Road,Elk Grove Village, Illinois 60007, (312) 593-2111.

"Technology Management by Robert Goldscheider," Clark Boardman Company, Ltd., 435 Hudson Street, New York, New York 10014,(1-800-221-9428).

"Valuation of Intellectual Property and Intangible Assets," Gordon V. Smith and Russell L. Parr, John Wiley & Sons, Inc.,605 Third Avenue, New York, New York 10158-0012.

"The New Alliance: America's R&D Consortia," Dan Dimncescu and James Botkin, Ballinger Publishing Company, 54 Chruch Street, Harvard Square, Cambridge, Massachusetts 02138, 1-800-638-3030.

"Technology and the American Economic Transition: Choices for the Future," Congress of the United States, Office of Technology Assessment, Washington, D.C. 20510-8025.

"Joint Ventures and Corporate Partnerships," Jennifer Lindsey,Probus Publishing Company, 118 North Clinton Street, Chicago, IL 60606, 1-800-426-1520.

"Managing For Joint Venture Success," Kathryn Rudie HarriganLexington Books, D.C. Heath & Company, 125 Spring Street, Lexington, Massachusetts 02173.

"Who Owns Innovation?" Robert A. spanner, Dow Jones-Irwin, Homewood, Illinois 60430.

Training

Managing Technology as A Strategic Resource, California Institute of Technology, Industrial Relations Center 1-90, Pasadena, California 91125, (818) 356-4041.

Patents, Trade Secrets and Licensing of Technology, Franklin Pierce Law Center, 2 White Street, Concord, New Hampshire 03301,(603) 228-1541.

International Licensing and Negotiation for the Technology Manager, Center for Professional Advancement, P.O. Box H East Brunswick, New Jersey 08816-0257, (201) 613-4500.

Appendix A

Licensing and Acquiring New Products for Technology American Management Association, P.O. Box 319, Saranac Lake, New York 12983, (518) 891-0065.

APPENDIX B

LEGISLATION

- 37 Code of Federal Regulations 401: Rights to Inventions Made by Nonprofit Organizations and Small Business Firms Under Government Contracts and Cooperative Agreements
- 37 Code of Federal Regulations 404: Licensing of Government-Owned Inventions
- Public Law 96-517, Bayh-Dole Act of 1980
- Public Law 99-502, The Federal Technology Transfer Act of 1986
- Public Law 100-418, Omnibus Trade and Competitiveness Act of 1988
- Public Law 96-480, Stevenson-Wydler Innovation Act of 1980
- Cooperative Research Act of 1984

APPENDIX C

SAMPLE AGREEMENTS:
MODEL COOPERATIVE RESEARCH AND DEVELOPMENT AGREEMENT
BETWEEN _____ AND THE NATIONAL BUREAU OF STANDARDS ON A RESEARCH ASSOCIATE PROGRAM

The National Bureau of Standards, hereinafter referred to as NBS, agrees to supervise and administer on behalf of _____, hereinafter referred to as the Sponsor, a Research Associate Program relating to _____ The parties agree as follows:

ARTICLE 1
DEFINITIONS

1.1 The term "cooperative research program" means the research activities described in Article 2 that are jointly undertaken by NBS and one or more non-Federal parties that have entered into a Cooperative Research and Development Agreement with NBS for that purpose.

1.2 The term "invention" means any invention or discovery which is or may be patentable under Title 35 of the United States Code or any novel variety of plant which is or may be protectable under the Plant Variety Protection Act (7 U.S.C. 7321 et seq.).

1.3 The term "made" in relation to any invention means the conception or first actual reduction to practice of such invention.

1.4 The term "Proprietary Information" means information which embodies trade secrets developed at private expense or which is confidential business or financial information provided that such information:

(i) It is not generally known or available from other sources without obligations concerning its confidentiality;

(ii) Has not been made available by the owners to others without obligation concerning its confidentiality; and

(iii) Is not already available to the Government without obligation concerning its confidentiality.

1.5 The term "Subject Data" means all recorded information first produced in the performance of this Agreement.

1.6 The term "Subject Invention" means any invention conceived or first actually reduced to practice in the performance of work under this Agreement.

ARTICLE 2
STATEMENT OF WORK

Cooperative research performed under this Agreement shall be performed in accordance with the Statement of Work attached hereto as Appendix A. Any modification to this initial scope shall be by mutual agreement between the Sponsor and NBS.

The following activities which complement this Program are being pursued at NBS, primarily in the _____

_____
(Name of the NBS Organizational Unit)

ARTICLE 3
PROGRAM DETAILS

3.1 The Program described in Appendix A shall be conducted during the period commencing on or about _____ and ending on or about _____, subject to extension by mutual agreement of the parties hereto, and to the provisions of sections 3.3 and 6.1.

3.2 NBS shall be the supervising agency, both administrative and scientific, for this Research Associate Program. _____
(Name of Individual)

Appendix C

_____
(Title)
_____
(NBS Organizational Title)
shall serve as NBS Supervisor.

3.3 The scientific and technical program of the Research Associate shall be reviewed at least quarterly by both parties of this Agreement and more frequently if deemed advisable by either party. Such review shall precede approval of the work program for each suceeding period. _____ shall represent the Sponsor in these reviews.

3.4 _____ shall serve as Research Associate for the Program described herein. The provisions applying to Research Associates as this term is used in this Agreement shall also apply to members of their supporting staff while serving at NBS as employees of the Sponsor. (See Note 1.)

3.5 While it shall be the privilege and responsibility of the Sponsor to select the Research Associates, they shall also be acceptable to NBS.

3.6 The Sponsor shall reimburse NBS for the cost of special supplies, special material, computation, technical assistance, and/or other special services provided the Research Associate(s) by NBS in connection with the Program covered by this Agreement. (The Sponsor shall, as of the starting date of the period covered by this Agreement, establish a fund of $_____ at NBS from which such reimbursement may be drawn. See Note 2.)

3.7 Charges for such special supplies and/or services shall require the approval of the Research Associate and the NBS Supervisor for this Program. (See Note 3.) Such charges shall not exceed $_____ during the period covered by this Agreement without prior approval of the Sponsor. (Upon termination of the Program, outstanding charges shall be deducted from the fund established with NBS by the Sponsor to cover such costs, and the remainder of such monies shall then revert to the Sponsor. (See Note 2.) All equipment, materials, instruments and supplies purchased during the term of this Agreement from funds in the account shall be and remain at all times the property of NBS).

3.8 Special equipment and instruments obtained by the Sponsor from sources external to NBS and provided by the Sponsor to NBS for use in connection with the Program covered by this Agreement will be returned to the Sponsor at the Sponsor's expense and risk as soon as practicable after termination of this Agreement. The Sponsor agrees to assume full responsibility for maintenance of such equipment and instruments and agrees to hold NBS free from liability for any loss thereof or damage thereto.

3.9 Remuneration to the Research Associate(s) for travel and related expenditures shall be provided directly by the Sponsor.

3.10 Research Associates shall pursue their activities at NBS on the work schedule and under the Government security and conduct regulations that apply to NBS employees. Research Associates shall conform to the requirements of Department of Commerce Administrative Orders 202-735 and 202-735-A, as amended, hereby made part of this Agreement, to the extent that these orders prohibit private business activity or interest incompatible with the best interests of the Department. (See Note 4.)

ARTICLE 4
PATENT RIGHTS

4.1 *Reporting.* NBS shall promptly report to the each Subject Invention disclosed to NBS by its employees. The Sponsor shall promptly report to NBS each Subject Invention disclosed to it by the Research Associate.

4.2 *Research Associate Inventions.* NBS, on behalf of the U.S. Government, waives any ownership rights the U.S. Government may have in Subject Inventions made solely by the Research Associate under the project and agrees that the Research Associate, or the Research Associate's Sponsor pursuant to any existing employment agreement, shall have the option to retain title to any such employee Subject Invention. If either the Sponsor or the Research Associate elects to take title to a Subject Invention under this section, that party

Appendix C

shall promptly notify NBS upon making the election and shall file patent applications on such Subject Invention at its own expense and in a timely fashion. Any party electing to take title to a Subject Invention under this section agrees to grant to the U.S. Government a nonexclusive, irrevocable, paid-up license in the patents covering a Subject Invention to practice the invention, or to have it practiced, throughout the world by or on behalf of the U.S. Government. Such nonexclusive license shall be evidenced by a confirmatory license agreement prepared by the in a form satisfactory to NBS. (See 4.4)

4.3 *NBS Employee Inventions.* NBS, on behalf of the U.S. Government, shall have the initial option to retain title to each Subject Invention made by its employees and in each Subject Invention made jointly by a Research Associate and an NBS employee. In the event that the NBS informs the Sponsor that it elects to retain title to such joint Subject Invention, the Sponsor agrees to assign whatever right, title and interest it has in and to such joint Subject Invention. NBS may release the rights provided for by this paragraph to employee inventors or to the subject to a license in NBS. (See 4.4)

4.4 *Filing of Patent Applications.* The party having the right to retain title and file patent applications on a specific Subject Invention may elect not to file patent applications thereon provided it so advises the other party within 90 days from the date it reports the Subject Invention to the other party. Thereafter, the other party may elect to file patent applications on such Subject Invention and the party initially reporting such Subject Invention agrees to assign its right title and interest in such Subject Invention to the other party and cooperate with such party in the preparation and filing of patent applications thereon. The assignment of the entire right title and interest to the other party pursuant to this paragraph shall be subject to the retention by the party assigning title of a nonexclusive, irrevocable, paid-up license to practice, or have practiced, the Subject Invention throughout the world. In the event neither of the parties to this agreement elect to file a patent application on subject invention, either or both (if a joint invention) may, at their sole discretion and subject to reasonable conditions, release the right to file to the inventor(s) with a license in each party of the same scope as set forth in the immediate preceding sentence.

4.5 *Patent Expenses.* All of the expenses attendant to the filing of patent applications as specified in 4.4 above, shall be borne by the party filing the patent application. Any post filing and post patent fees shall also be borne by the same party. Each party shall provide the other party with copies of the patent applications it files on any Subject Invention along with the power to inspect and make copies of all documents retained in the official patent application files by the applicable patent office.

4.6 *Exclusive License*

4.6.1 *Grants.* NBS, on behalf of the Government, hereby agrees to grant to the Sponsor an exclusive license in each U.S. patent application, and patents issued thereon, covering a Subject Invention, which is filed by NBS on behalf of the U.S. Government subject to the reservation of an irrevocable, royalty-free license to practice and have practiced the Subject Invention on behalf of the U.S. Government, and such other terms and conditions as are specified by NBS in such exclusive license.

4.6.2 *Exclusive License Terms.* Upon filing of a patent application on a Subject Invention by NBS, the Sponsor shall have the option to acquire an exclusive license in the resulting patents at reasonable royalty rates upon the execution of an exclusive license agreement. The specific royalty rate shall be negotiated promptly after the Subject Invention is filed in the U.S. Patent and Trademark Office, provided that this option must be exercised by the Sponsor by written notice to NBS within three months from the date the U.S. Patent Application is so filed. The reasonable royalty rate for each exclusive license shall be based upon a portion of the selling price of the item attributable to the presence of claimed subject matter where such item is a machine, article of manufacture, product made by a process, or composition of matter as defined by the claims of the patents. Where the claimed subject matter relates to a process or method to be practiced under the claims of the patent, the royalty will be based upon the net savings attributable to the implementation of said process or method.

ARTICLE 5
DATA AND PUBLICATION

5.1 *Release Restrictions.* NBS shall have the right to use all Subject Data for any Governmental purpose, but shall not release such Subject Data publicly except: (i) NBS when reporting on the results of sponsored research may publish Subject Data, subject to the provisions of paragraph 5.3 below; and (ii) NBS may release such Subject Data where such release is required pursuant to a request under the Freedom of Information Act (5 U.S.C. Section 552); provided, however, that such data shall not be released to the public if a patent application is to be filed (35 U.S.C. Section 205) until the party having the right to file has had a reasonable time to file.

5.2 *Proprietary Information.* The Sponsor shall place a Proprietary notice on all information it delivers to NBS under the Agreement which the Sponsor asserts is proprietary. NBS agrees that any information designated as proprietary which is furnished by the Sponsor to NBS under this Agreement, or in contemplation of this Agreement, shall be used by NBS only for the purpose of carrying out this Agreement. Information designated as proprietary shall not be disclosed, copied, reproduced or otherwise made available in any form whatsoever to any other person, firm, corporation, partnership, association or other entity without the consent of the Sponsor except as such information may be subject to disclosure under the Freedom of Information Act (5 U.S.C. 552). NBS agrees to use its best efforts to protect information designated as proprietary from unauthorized disclosure. The Sponsor agrees that NBS is not liable for the disclosure of information designated as proprietary which, after notice to and consultation with the Sponsor, NBS determines may not lawfully be withheld or which a court of competent jurisdiction requires disclosed.

5.3 *Publication*

5.3.1 Work completed by Research Associates shall be made available to the public under the same conditions as work performed by NBS employees. In no event, however, shall the name of the Sponsor or any of its trademarks and tradenames be used in NBS publications without its prior written consent.

5.3.2 NBS and the Sponsor agree to confer and consult prior to the publication of Subject Data to assure that no Proprietary Information is released and that patent rights are not jeopardized. Prior to submitting a manuscript for review which contains the results of the research under this Agreement, or prior to publication if no such review is made, each party shall be offered an opportunity to review such proposed publication and to file patent applications in a timely manner, if it is so entitled under this Agreement.

ARTICLE 6
TERMINATION

6.1 The Sponsor and NBS each have the right to terminate this agreement, or the association with any individual Research Associate, upon 60 days notice in writing to the other party.

6.2 In the event of termination by NBS, NBS shall repay the Sponsor any prorated portion of payments previously made to NBS pursuant to Article 3 of this Agreement in excess of actual costs incurred by NBS in pursuing this project. A report on results to date of termination will be prepared by NBS and the cost of the report wil be deducted from any amounts due to the Sponsor from NBS.

ARTICLE 7
DISPUTES

7.1 *Settlement.* Any dispute arising under this Agreement which is not disposed of by agreement of the parties shall be submitted jointly to the signatories of this Agreement. A joint decision of the signatories or their designees shall be the disposition of such dispute.

7.2 If the signatories are unable to jointly resolve a dispute within a reasonable period of time after submission of the dispute for resolution, the matter shall be submitted to the Director of NBS or his designee for resolution.

7.3 *Continuation of Work.* Pending the resolution of any dispute or claim pursuant to this Article, the parties agree that performance of all obligations shall be pursued diligently in accordance with the direction of the NBS signatory.

Appendix C

ARTICLE 8
LIABILITY

8.1 *Property.* The U.S. Government shall not be reponsible for damages to any property of the Sponsor provided to NBS or acquired by NBS pursuant to this Agreement.

8.2 *Indemnification.*

8.2.1 Research Associates are not employees of NBS. The Sponsor and the Research Associate agree to indemnify and hold harmless the U.S. Government for any loss, claim, damage, or liability of any kind involving the Research Associate arising in connection with this Agreement, except to the extent that such loss, claim, damage or liability arises from the negligence of NBS or its employees. NBS shall be solely responsible for the payment of all claims for the loss of property, personal injury or death, or otherwise arising out of any negligent act or omission of its employees in connection with the performance of work under this Agreement.

8.2.2 The Sponsor holds the U.S. Government harmless and indemnifies the Government for all liabilities, demands, damages, expenses and losses arising out of the use by the Sponsor, or any party acting on its behalf or under its authorization, of NBS's research and technical developments or out of any use, sale or other disposition by the or others acting on its behalf or with its authorization, of products made by the use of NBS' technical developments. This provision shall survive termination of this Agreement.

8.3 *Force Majeure.* Neither party shall be liable for any unforeseeable event beyond its reasonable control not caused by the fault or negligence of such party, which causes such party to be unable to perform its obligations under this Agreement (and which it has been unable to overcome by the exercise of due diligence), including, but not limited to, flood, drought, earthquake, storm, fire, pestilence, lightning and other natural catastrophes, epidemic, war, riot, civic disturbance or disobedience, strikes, labor dispute, or failure, threat of failure, or sabotage of the NBS facilities, or any order or injunction made by a court or public agency. In the event of the occurrence of such a force majeure event, the party unable to perform shall promptly notify the other party. It shall further use its best efforts to resume performance as quickly as possible and shall suspend performance only for such period of time as is necessary as a result of the force majeure event.

ARTICLE 9
MISCELLANEOUS

9.1 *Governing Law.* The construction validity, performance and effect of this Agreement for all purposes shall be governed by the laws applicable to the Government of the United States.

9.2 *Amendments.* If either party desires a modification in this Agreement, the parties shall, upon reasonable notice of the proposed modification by the party desiring the change, confer in good faith to determine the desirability of such modification. Such modification shall not be effective until a written amendment is signed by all the parties hereto by their representatives duly authorized to execute such amendment.

9.3 *Notices.* All notices pertaining to or required by this Agreement shall be in writing and shall be directed to the signator(s).

9.4 *Independent Contractors.* The relationship of the parties to this Agreement is that of independent contractors and not as agents of each other or as joint venturers or partners. NBS shall maintain sole and exclusive control over its personnel and operations.

9.5 *Use of Name or Endorsements.* (a) the Research Associate or Sponsor shall not use the name of NBS or the Department of Commerce on any product or service which is directly or indirectly related to either this Agreement or any patent license or assignment agreement which implements this Agreement without the prior approval of NBS. (b) By entering into this Agreement NBS does not directly or indirectly endorse any product or service provided, or to be provided, by the Sponsor its successors, assignees, or licensees. The Sponsor shall not in any way imply that this Agreement is an endorsement of any such product or service.

ARTICLE 10
DURATION OF AGREEMENT AND EFFECTIVE DATE

10.1 *Duration of Agreement.* The Program described in Appendix A shall be conducted during the period commencing on or about _____ and ending on or about _____, subject to extension by mutual agreement of the parties hereto.

10.2 *Effective Date.* This Agreement shall enter into force as of the date of the last signature of the parties.

IN WITNESS WHEREOF, the Parties have caused this Agreement to be executed by their duly authorized representatives as follows:

Signed for the Sponsor: _____

_____
(Title)

Signed By the Research Associate: _____

_____
(Title)

_____
(Date)

For the National Bureau of Standards: _____
DEPUTY CHIEF COUNSEL

_____
(Date)

_____
MOU DIRECTOR

_____
(Date)

_____
DIRECTOR, ORTA

EXCLUSIVE LICENSE AGREEMENT

UNIVERSITY MODEL TECHNOLOGY TRANSFER AGREEMENT

On the broader spectrum of a university-sector technology transfer involving a private-sector individual or corporate entity, the business incubator is a significant facilitator of transfers by networking technical talent, technology, capital and professional expertise. The university-private sector incubator also can shorten the inventor's learning curve, decrease development and commercialization timeframes and, ultimately, solve a marketplace problem more expeditiously.

Recitals

1. Whereas, sponsor desires to retain university to do certain work under the direction of principal investigator and Whereas, university desires to undertake the performance of the above under the terms and conditions hereinafter set forth and; Now, therefore, in consideration of the mutual promises and covenants described herein, the parties hereto, intending to be legally bound hereby, agree as follows:

2. Whereas university desires to obtain support for research in the field of ____ (hereinafter said research in said field is called "Research") and Whereas, sponsor desires to support (research) and obtain rights to any know-how and patent rights resulting from (research), Now, therefore:

3. Whereas university has applied to sponsor for, and sponsor has agreed to furnish, funds for the support of a research and study project which is described in a proposal submitted to sponsor by university titled ____, a copy of which is attached and which is hereafter referred to as the "Research Project" and made a part of this agreement; and such additional projects as are subsequently agreed upon by the sponsor and the university in writing. Now, therefore:

4. Whereas, university has represented that it is equipped and qualified to perform research and development necessary to evaluate, and; Whereas, sponsor desires to support the studies

proposed by university in the proposal dated_____, and to obtain the results thereof on behalf of (prime contractor); Now, therefore, the parties agree that university shall furnish the materials, facilities, equipment, personnel, services, and all other necessary items for the performance of a program of research, all as more fully set forth in the following Attachments to this agreement:

5. Whereas, sponsor and its subsidiaries are engaged in the manufacture and/or sale of (products), and in connection therewith are engaged in the business of research and development and of making various inventions, discoveries, improvements and developments in various scientific fields, and more specifically in the fields of _____, and in the exploitation of development of processes, designs, apparatus and devices for use in said fields; Now, therefore:

6. Whereas, sponsor needs to obtain information on _____ and needs to develop new methods for the detection of _____ and whereas (technique) is being widely applied to a variety of problems at university and university is interested in carrying out additional studies in _____. Sponsor and university do hereby undertake a joint effort to obtain data for _____ and develop new methods for _____.

7. Whereas, university and sponsor have in common the desire to encourage and facilitate the discovery, synthesis, application and dissemination of new knowledge of _____; and whereas, this common desire can be fulfilled by participation in the collaborative research project described below, and Whereas it is the intent of this agreement to provide maximum scientific and administrative flexibility in the conduct of this project; now, therefore:

8. Whereas, the university has facilities and a professional staff under the direction of _____ for conducting research in the field of _____ by the use of _____ techniques as described in the Research Proposal. The research to be carried out by the university in its department of _____ at the _____ campus in _____ under the terms and conditions specified herein; and

Whereas the (sponsor) desires to fund said research under the terms and conditions specified herein, and to acquire an exclusive, worldwide license from university for any technology resulting from said research in accordance with the terms and provisions of the License Agreement set forth in _____; Now, therefore, in consideration of the foregoing, and in consideration of the mutual covenants, conditions, and undertakings hereinafter set forth, the parties do hereby covenant and agree as follows:

9. Whereas, the (sponsor) desires research services in accordance with the scope of work outlined within this agreement, and Whereas the performance of such research is consistent, compatible and beneficial to the academic role and mission of the (university) as an institution of higher education and, in consideration of the mutual promises and covenants contained herein, the parties hereto agree as follows:

10. Whereas, the _____ department of (corporation) seeks to improve the _____ services it provides to its users through a better understanding of alternative approaches to scientific _____, and Whereas, (university) desires to improve the operation of its _____ through further study and development of (named) system, Now, therefore, the parties agree as follows:

Description of Research and Reporting

1. University shall provide the necessary personnel, equipment, facilities, and supplies to perform the services specified in the attached Statement of Work, marked Exhibit A, which by this reference is made a part hereof for all purposes.

2. University shall submit technical reports of work accomplished to the sponsor's project director in the form and in accordance with the schedule contained in Exhibit A. If so requested by the sponsor's project director, university hereby agrees to revise any or all of these reports in accordance with instructions from the sponsor's project director. The research program will be conducted in accordance with the proposal by _____ attached hereto as Appendix A. This program may be

Appendix C

revised from time to time by mutual agreement as the research proceeds. During the term of the research program, ____ will submit semiannual written reports to sponsor, setting forth the technical progress made during the previous six month period and identifying the research effort and goals to be undertaken during the next six month period.

The Research and Development Project

1. A research and development project is hereby established at university as provided below for the purpose of conducting research, enquiries, investigations and studies during the term of this agreement relating to ____ as described in ____ dated ____ to the sponsor from university (which research, enquiries, investigations and studies are herein collectively referred to as the "Project").

2. University shall provide as set out below its services and as much time and effort as are reasonably needed for the proper conduct of the Project, within the limits of the payments to be made by the sponsor.

Progress of the Work, Reports and Communications

1. University shall periodically, and not less frequently than once every two months, during the term of this agreement report in writing to the sponsor on the progress of the Project and the results being obtained and shall make available to the sponsor all information relative thereto. In particular such reports shall fully and promptly disclose to the sponsor any formulae, inventions, discoveries and improvements made or discovered by university in the course of the Project, whether or not patentable.

Description of the Work

1. University shall furnish all necessary labor, materials, and facilities and shall exert its best efforts in carrying out the

specific objectives set out below in support of university's grant no. ____.

2. Research conducted under this agreement will be planned cooperatively by the university and sponsor. It is understood, however, that ____ as principal investigator, has the responsibility for the scientific and technical conduct of this project.

3. University shall render reports as necessary for annual progress reports and for a terminal progress report. The schedule of these reports will be coordinated with the principal investigator.

Modifications

The sponsor and the university agree that this agreement may be modified or changed by mutual agreement. Such modifications or changes so executed shall be in writing, signed by the original signatories or their successors, assignees or agents, and shall be attached to and become part of the agreement.

Level of Effort

____ agrees to devote approximately ____ percent Full Time Equivalent to collaborate in this project. University agrees to hire and assign appropriate technical personnel for this project. No funds other than those requested in the approved budget will be required for collaboration.

Definitions

For the purpose of this agreement, the following terms shall have the following meanings:

Authorized Research Project—The term refers to a research project to be designated and defined, both as to scope and objective, by sponsor and performed by university's principal investigator in the general area of ____, the cost of which, at university's usual rates, shall be agreed upon between the parties prior to initiation of any work thereunder and paid by sponsor.

Appendix C

Completion

It is understood that the nature of this research project is such that the university does not guarantee the completion of the project within the project period or within the limits of the financial support hereby furnished.

Sponsor's Obligation

Sponsor agrees to expend, at a minimum ____ dollars on authorized research projects during the ____ years following the date of this agreement. For the purpose of this paragraph, research projects which have been defined and agreed upon between the parties during that ____ year period shall be considered as having been fully performed during that ____ year period regardless of the date the project is actually completed, and final payment with respect to that project otherwise becomes due and payable. In the event the parties shall fail to reach agreement as to the scope, definition and cost of sufficient authorized research projects so that sponsor may meet this obligation, sponsor may elect to pay the balance of the financial commitment hereby undertaken outright to university, and shall then be deemed to have satisfied all conditions of this paragraph.

University's Obligations

In return for the promises of sponsor as set forth in Section ____, university hereby undertakes the following obligations to sponsor:

1. *Grant of License*—University hereby grants the license set forth in section ____ of this agreement to sponsor and its majority-owned affiliated companies. The physical embodiments of the ____, shall be delivered to sponsor herewith.
2. *Research Projects*—University further agrees to perform the authorized research projects required hereunder, as such may be defined by the parties.
3. *Research Product*—With respect to rights in any software developed in the performance of authorized research projects, university agrees as follows:

a. That if the purpose or object of a particular ____ is defined for university by sponsor and developed by university as the object of an authorized research project, sponsor shall be granted an exclusive, worldwide and perpetual license to use, copy and sell ____ without payment of royalty or fee, subject only to the right of university to use such ____ for its own, internal purposes. If requested, university agrees to transfer or assign all copyrights in and to such ____ to sponsor.

b. That if the purpose or object of the particular ____ is defined by university and developed by university during the course of performance of authorized research projects, sponsor shall be granted a nonexclusive, worldwide and perpetual license to use and copy such software without payment of royalty or fee of any sort.

c. In either event a. or b. above, if ____ is incorporated in such ____, no further royalty or license fee shall be charged to sponsor with respect to ____ except as provided in section ____.

4. *Enhancements*—University agrees to provide sponsor, without fee or royalty, any enhancements or improvements to the basic which it develops or incorporates in such package during the ____ year period referred to in paragraph ____. Thereafter for a period of years, university agrees to make such enhancements or improvements available to sponsor on the same terms and conditions available to other Licensees of ____.

Supervision

1. The research will be supervised by ____. If for any reason he is unable to continue to serve as principal investigator, and a successor acceptable to both university and sponsor is not available, this agreement shall be terminated as provided in termination clauses.

2. The conduct of the investigation shall be under the full control of the university, which will supervise and direct all experi-

mental work and the computation and reduction of all results obtained, together with the placing of the resulting data into form for presentation, substitutions for the key personnel shall be made by the university only upon the prior written approval of sponsor. If at any time the principal investigator determines that the research data dictate a substantial change in the direction/focus of the investigation, he will promptly notify the sponsor of such information. In this event, the university and sponsor shall make a good faith effort to agree on any necessary changes in the scope of the investigation. If agreement is not reached, either the university or the sponsor may terminate this agreement upon not less than ninety days written notice to the other party.

3. Performance of the work shall be supervised by the university under the general monitoring of a project manager as designated by sponsor. Sponsor may, at any time, designate a new or alternate project manager upon written notification to the university and sponsor may, at any time, by written order to the university, make changes within the general scope of the work, including but not limited to revising or adding to the work or deleting portions thereof, or revising the period of schedule or performance. The individuals set forth in the proposal are considered essential to the work being performed under this agreement; substitutions for any such individuals or substantial reductions in any of their efforts require sponsor approval.

4. University's relationship to sponsor in the performance of this agreement is that of independent contractor. The personnel performing services under this agreement shall at all times be under university's exclusive direction and control and shall be employees of university and not employees of the sponsor.

5. It is understood and agreed that university is an independent contractor in the performance of each and every part of this contract and that contractor's employees shall be the employees of contractor and shall be subject to contractor's sole and exclusive supervision, direction and control. Sponsor shall

have the right to inspect and direct the performance of the services to ensure satisfactory completion thereof, it being distinctly understood that sponsor is in no way associated or otherwise connected with the actual performance and the details of the services to be performed in connection with this contract as sponsor is interested in and looking only to the end result to be accomplished, and that contractor is solely liable for all labor and expenses in connection therewith. Contractor shall not assign this contract or subcontract the whole or any part of the services to be performed by contractor hereunder without the sponsor's prior written consent. Sponsor's consent to any such assignment or subcontract shall not relieve contractor or its surety if there is a surety, of any liability for the full and faithful performance of this contract according to its terms and conditions.

6. University agrees to permit sponsor's representatives to confer, from time to time, with university's representatives both personally and by telephone and to witness performance of the work hereunder. It is understood and agreed that sponsor's representatives have no authority to supervise, direct or control university's representatives and that in all respects, the carrying out of the work shall be under the university's supervision and control.

7. The areas to be included in the project shall be determined by mutual agreement between the parties. The methods employed in the field and office shall be those adopted by sponsor to ensure the required standards of accuracy, provided, however, that in absence of applicable sponsor standards or in cases of conflict between sponsor standards and other standards that would apply, the parties shall determine the applicable standards by mutual agreement.

8. The professional services of contractor in connection with the work to be performed shall be under the general direction of sponsor's director. Such services shall be limited to those described within the scope of this agreement. All contractual matters relating to this agreement are to be referred to

Appendix C

sponsor's subcontract administrator, who is the only person authorized to alter any provisions of this agreement on behalf of sponsor.

9. Regular conferences between the university and the sponsor shall be held for the purpose of discussing the progress of the project, the results being obtained and the areas and directions in which further work on the Project is to be conducted. University shall conduct the work on the project during the term of this agreement in such areas and directions as shall be designated from time to time by the sponsor so long as such areas and directions are acceptable to university. Sponsor may decide against acceptance for any reason, e.g., the inherent nature of the work or other work then being or previously done by university.

10. For the purposes of this agreement and all services to be provided hereunder, university shall be, and shall be deemed to be, an independent contractor and not sponsor's agent or employee. University shall have no authority to make any statements, representations, or commitments of any kind, or to take any action which shall be binding on sponsor, except as provided for herein or authorized in writing by sponsor.

11. It is understood that in the performance of this agreement university is acting solely as an independent contractor and not as an employee of sponsor. Further, nothing in this agreement shall be construed or applied to create a relationship of partners, agency, joint adventurers or of employer and employee. As university is an independent contractor, it is understood that sponsor has no obligations under state or federal laws regarding employee liability and that the total commitment and liability of sponsor in regard to this arrangement is fee limited as described above.

12. Contractor's relationship to sponsor under this agreement will be that of an independent contractor. Contractor is to exercise its own discretion on the method and manner of performing its duties and sponsor will not exercise control over contractor or its employees except regarding the result to be obtained and

to assure compliance with the terms of this agreement. Sponsor reserves the right to offer suggestions to contractor regarding the technical aspects of contractor's services. Personnel retained or assigned by contractor to perform services under this agreement will at all times be considered employees of contractor and not as employees or agents of sponsor.

13. This agreement sets forth the terms of a joint project between corporation and university, who are designated as managing coordinators for this joint project. It shall be the duty of the coordinators to mutually select research topics, arrange to transmit and receive information from the other party, coordinate visits and arrange all other matters pertinent to this effort. Corporation will assign _____ to university for _____ time. During this period, _____ will remain a regular employee of corporation and _____ will, in collaboration with _____ and other staff of university, investigate _____. University will provide _____ with office and _____ appropriate for his investigations, as well as selected information on _____.

14. Sponsor personnel will be given reasonable access to the research records and facilities at the university relating to the research program and will have the right to participate with _____ in the research program within the bounds of open scientific exchange between colleagues, it being understood that in the event any formal training or work program is required by sponsor, specific arrangements for such program will be made with the university. Sponsor will coordinate contact between its personnel and _____ - to avoid unnecessary and repetitive visits and disruptions to the research program. _____ and his colleagues in the research program will be given access to sponsor facilities relating to the research program, subject, however, to reasonable restrictions by sponsor to protect proprietary information not directly relevant to the success of the research program.

15. Regular conferences between university and the client shall be held for the purpose of discussing the progress of the project, the results being obtained and the areas and directions in

which further work on the project is to be conducted. University shall conduct the work on the project during the term of this agreement in such areas and directions as shall be designated from time to time by the sponsor so long as such areas and directions are acceptable to university. Sponsor may decide against acceptance for any reason, e.g., the inherent nature of the work or other work then being or previously done by university.

16. *Independent inquiry.* Nothing in this agreement shall be construed to limit the freedom of researchers who are not participants in this agreement, whether paid under this agreement or not, from engaging in similar research inquiries made independently under other grants, contracts or agreements with parties other than sponsor.

Period of Agreement

Financial (Sample Clauses)

1. Upon acceptance of this agreement by both parties, sponsor agrees to pay university an initial amount equal to 40 percent of total project costs and 20 percent of total project costs at the beginning of subsequent three month periods.
2. Fiscal reporting under this agreement shall be considered complete by payment by the sponsor of invoices submitted.
3. Fiscal reporting under this agreement shall be provided to the sponsor within 90 days of the termination date. Funds not expended shall be returned at such time as this final accounting occurs.
4. No fiscal reporting of the expenditures made under this agreement is necessary to the sponsor.
5. Funding for the research program will be exclusively by sponsor unless additional funds are made available to the university for its unrestricted use by the U.S. Government or private sources which are approved by sponsor. The university will consult with sponsor regarding the use of any equipment

or facility in connection with the research program which has been acquired, in whole or in part, through U.S. Government funding. OMB Circular _____ is referenced herein as establishing the U.S. Government patent policy applicable to any government funding of the research program.

6. The university reserves the right to reallocate funds between approved budget categories. Sponsor approval, however, shall be obtained for transfer to categories not approved in the original budget.

7. Funds advanced under this agreement shall be subject to refund if the university or sponsor is unable to perform the terms of this agreement. The amount of the refund shall be subject to agreement by both parties to cover necessary costs and obligations to date or termination.

8. Requests for funds beyond those specified in this agreement shall pass through the institution's regular administrative process and shall be considered supplemental to the total amount currently authorized.

9. Expenditure limitation. The university shall not be obliged to spend any funds on this project other than those provided under this agreement. The university agrees to furnish the sponsor reports of the progress of the above described research during the term of this agreement.

Equipment (Sample Clauses)

1. All equipment purchased for use in connection with research project shall be the property of the university, provided that it shall be dedicated to that project while research is in progress.

2. The university agrees that all materials developed or acquired by the university under the auspices of this agreement shall become the property of the sponsor and the university shall, upon request by the sponsor, execute any such documents as may be necessary to effectively transfer such property under this provision.

Appendix C

3. The equipment listed in schedule _____ shall be provided by sponsor to university for the duration of this agreement. Title shall vest in sponsor. Costs of transportation and installation of the equipment shall be allowable under this agreement. University agrees to expend reasonable care in the use and storage of the equipment referred to in paragraph above, to maintain labels indicating that said equipment is the property of sponsor, and to return it to sponsor at the conclusion of all authorized research projects, unless otherwise amended.

4. Sponsor agrees to deliver to university for its use, for so long as authorized research projects are ongoing, _____, which equipment remain the property of sponsor. Delivery of said equipment shall be made within _____ weeks of the date of this agreement. Sponsor further agrees to pay all cost of maintaining and servicing said equipment so long as university is entitled, in accordance with this agreement, to keep said equipment in its possession.

Use of Names

Publication

1. Publications and copyrights. The university will be free to publish the results of research under this agreement, after providing a copy of the proposed publication to the sponsor. Title to and the right to determine the disposition of any copyrights or copyrightable materials, first produced or composed in the performance of this research, shall remain with the university, provided that the university shall grant to the sponsor an irrevocable, royalty-free, nonexclusive right to reproduce, translate and use all such copyrighted material for its own purposes.

2. Any reports and all proprietary information shall become and remain the property of the company, but the university retains the right to publish any scientific information in an appropriate scientific journal, subject to the following conditions:

a. The university shall not publish or have published any work during the performance of the contract or for a period of three months after completion of the project without the prior written consent of the company.

b. Upon expiration of the aforesaid period of three months, if the university intends to cause such scientific information to be published and to acknowledge that the work was performed under contract to the company, it shall give written notice of its intention to the company. Within 10 days of the receipt of such notice from the university, the company may, by notice in writing to the university, require the university to omit such reference.

3. University shall have the right to publish the research results of this study but shall provide company with a copy of any proposed paper before submittal for publication. University agrees to give good faith consideration to any comments or suggestions offered by company. If university decides not to publish the research results, company shall have the right to publish them. In order that publication of the results will not adversely affect company patent interest, company shall have the right to require university to delay submitting said paper for publication for a period of three months from the time the proposed paper is submitted to company for review. In such event, company will notify university in writing, of its desire to delay submission within thirty days of receipt of the proposed paper from university. Except insofar as the results of this study may be published under this paragraph, university agrees to take all reasonable steps to ensure that the work performed under this study remains confidential.

4. Results of the research herein outlined may be published jointly by the university and sponsor, or by either of these institutions separately, always giving due credit to the other and recognizing within proper limits the rights of the individuals doing the work. Manuscripts prepared for publication by either shall be submitted to the other party for suggestions and approval prior to publication. In the event of disagreement,

Appendix C

either party may publish results on its own particular activity giving proper acknowledgements of cooperation.

5. It is recognized that during the course of the work under this agreement, the university or its employees may from time to time desire to publish information regarding scientific or technical developments made or conceived in the course of or under this agreement. In order that premature public disclosure of such information not adversely affect the patent interests of company or the university, patent approvals for release and publication of such information shall be secured from the company project manager prior to any such release or publication, which approval shall not be unreasonably withheld.

6. Publication. The university shall have the exclusive right to publish or otherwise disclose at its discretion the results of the investigation. It will, however, inform the sponsor within a reasonable time of all such activity. In publishing or otherwise disclosing said material, the university will take all reasonable measures, including delay of publications for not longer than one year, in consultation with the sponsor, to ensure that the filing for patent rights in the United States or foreign countries is not thereby prohibited or jeopardized. Any publication or disclosure by the university shall give full credit to the sponsor and the university.

7. Publication and confidentiality. University, as a state institution of higher education, engages only in research that is compatible, consistent and beneficial to its academic role and mission and therefore significant results of research activities must be reasonably available for publication. The university agrees, however, to defer publication for a period not to exceed six months following completion of the project, unless it obtains sponsor approval prior to publication, which approval will not be unreasonably withheld by sponsor. University agrees to keep confidential any sponsor proprietary information supplied to it by sponsor during the course of research performed by the university, and such information will not be

included in any published material without prior approval by sponsor.

8. Publications and copyrights. The university will be free to publish the results of research under this agreement, after providing a copy of the publication to the sponsor. Title to and the right to determine the disposition of any copyrightable material, first produced or composed in the performance of this research, shall remain with the university provided that the university shall grant to the sponsor an irrevocable, royalty-free, nonexclusive right to reproduce, translate and use any such copyrighted material for its own purposes.

9. Publication. The sole and exclusive right of publication of this research investigation and the results thereof is hereby reserved to and shall remain the property of the university, but any such publication shall be delayed up to six months from the date university submits its request for publication to sponsor. The sponsor shall be provided with a draft of any proposed publication and shall respond within thirty days in order to request a delay in publication. Failure to respond within thirty days shall constitute de facto permission to publish. If university decides not to publish this research investigation and the results thereof, sponsor shall have the right to do so. If sponsor decides to publish, it shall allow the university to review the publication and give credit to the university and the principal investigator. The subject matter of any such publication shall not contain confidential information of the sponsor not already in the public domain and shall be confined to a statement of new discoveries and interpretations of scientific fact. The university shall have the exclusive right to publish at its discretion the results of the investigation. Prior to such publication, no publicity shall be given to any of the results of the investigation except upon the recommendation or with the approval of the university and the sponsor, unless the scientific value of a discovery made during the source of the investigation be such that, in the judgment of the university, the public interest requires prompt release or publication thereof. In any such publication full credit shall be

Appendix C

given the sponsor and every person and agency having made a significant contribution to the results obtained.

Patents and Licensing

1. All improvements and inventions made or conceived in the course of this grant and any patents or other rights resulting therefrom shall be the property of the university. University recognizes that patents will be an important consideration in any agreement entered into between the parties. Accordingly, university agrees to take steps to file patent applications in the United States at its expense on all patentable developments.

2. University agrees to grant sponsor an option to acquire an exclusive worldwide license to inventions made and conceived in the course of this grant pursuant to the terms of _____. The option must be exercised within one year of the filing of a patent application. If sponsor decides to exercise its option within the prescribed time period, the parties agree to negotiate in good faith an agreement satisfactory to both parties. All such negotiations including the execution of an agreement shall be completed within one year of written notice to university of sponsor's exercise of said option. If said agreement between university and sponsor is not signed in final form before expiration of the one year period above, university shall be free to negotiate with other companies not a party to this agreement without further obligation to sponsor with the provision that it shall not enter into any agreement having more favorable terms than those offered sponsor.

3. All rights, title and interest in and to inventions or other intellectual property rights made or conceived in the course of the research described in this proposal is hereby vested in university. University intends to apply for patents in the United States and foreign countries covering said inventions. Should university choose not to apply for any such patents, sponsor shall have the option to apply for the patents in the name of the university. Further, if after applying for such patents, university elects to discontinue prosecution or main-

tenance of any such application or patent, sponsor has the option to continue prosecution of maintenance. The cost incurred by sponsor in obtaining and maintaining United States of foreign patents shall be deductible from any royalty which would be paid to university on the sales or products covered by such applications or patents. Sponsor shall have an option during a period of up to one year from the date of filing said patent applications to obtain a time-limited world exclusive license with the right to sublicense from university to make, use, sell, manufacture or have manufactured products covered by the patent applications and patents within the field of the proposed research insofar as university is able to grant such license. If sponsor exercises the option to obtain said license, university and sponsor agree to negotiate in good faith the terms and conditions of such license including appropriate royalty payments to university.

4. If an invention is made solely by researchers at university, university agrees to grant sponsor an option to acquire an exclusive worldwide license to any invention made or conceived in the course of this grant. Where the invention is made by university researchers at sponsor facilities or by both university and sponsor investigators, university agrees upon request to assign its right, title, and interest in and to any patents to sponsor. If sponsor decides to exercise its option to acquire an exclusive license or to request an assignment of any patents, the parties agree to negotiate in good faith an agreement satisfactory to both parties, taking into account the matters of inventorship, technology base and use of facilities in determining a fair and reasonable royalty.

5. University shall have the right to publish the results of the research falling within the scope of this grant, provided, however, that both publications and texts of oral presentations will be submitted to the university patent administrator and to sponsor for review. If the publication describes a potentially patentable development, submission for publication shall be withheld only so long as is necessary to file patent applications. Title to and the right to determine the disposition of any

copyrights or copyrightable material first produced or composed in the performance of this research shall remain with university, provided that university shall grant to sponsor an irrevocable, royalty-free, nonexclusive right to reproduce, translate and use all such copyrighted material for its own purposes. At sponsor's request, acknowledgement of the support received from sponsor will be made in all publications. University shall provide sponsor with periodic research reports of the proposed research at least semiannually.

6. While university shall retain the right to publish its own research results, university agrees to submit manuscripts of any proposed publication relating to the field of proposed research to sponsor for a period of at least 90 days for review prior to submission to any journal or other publication source. Upon written request from sponsor, university agrees to delay submission in order to permit university and/or sponsor to protect their proprietary interests. Title to and the right to determine the disposition of any copyrights or copyrightable material first produced or composed in the performance of this research shall remain with university, provided that university shall grant to sponsor an irrevocable, royalty-free, nonexclusive right to reproduce, translate and use all such copyrighted material for its own purposes.

7. Nothing in this agreement shall prevent university personnel from publishing the results of research conducted under this agreement, provided however, that university agrees that it will not permit such personnel to publish any results of such research without first providing sponsor with a reasonable opportunity to file United States patent applications on university's behalf seeking patent protection for patentable subject matter contained in such results and, provided further, that no publication of such results shall contain confidential or proprietary information obtained from sponsor by university personnel during the course of such research. Reasonable opportunity shall be no more than three months from the date on which sponsor has notice of such personnel's intent to

publish such results and sponsor has been provided with a manuscript or written report fully describing such results.

Liability

1. Indemnification. Sponsor hereby waives any claim against the university and agrees to indemnify, defend, and hold harmless university from any loss, claim, damage, or liability of any kind involving an employee of sponsor arising out of or in connection with this agreement, except to the extent that such loss, claim, damage, or liability arises in whole or in part from the negligence of university.

2. Warranties. University makes no warranties, express or implied, as to any matter whatsoever including, without limitation, the condition of the research or any invention or product, whether tangible or intangible, conceived, discovered, or developed under this agreement; or the ownership, merchantability, or fitness for a particular purpose of the research or any such invention or product. University shall not be liable for any direct, consequential, or other damages suffered by any licensee or any others resulting from the use of the research or any such invention or product.

3. *Limits of Liability*. University is and will be acting as an independent contractor in the performance of this work, and it shall be solely responsible where found liable to the extent covered by insurance for the payment of any and all claims for loss, personal injury, death, property damage, or otherwise, arising out of any act or omission of its employees or agents in connection with the performance of this work.

4. *Company's Right of Recovery*. Nothing in the above paragraphs shall be considered to preclude company from receiving the benefits of any insurance university may carry which provides for the indemnification of any loss or destruction of, or damages to property in the custody and care of university. University shall do nothing to company's right to recover against third parties for any loss, destruction of, or damage to

property, and upon the request of the company project manager shall, at company's expense, furnish all reasonable assistance and cooperation including assistance in the prosecution of suit and the execution of instruments of assignment in favor of company obtaining recovery. Company shall not be liable for any injury to or death of any person or persons or for loss to university personnel or damage to university property unless such injury or damage is due to negligence on the part of the company.

5. *Insurance.* University shall maintain in force at its sole cost and expense, with insurance companies acceptable to company, insurance policies of the types and in the minimum amounts and with the terms and conditions set forth in the Exhibit C attached hereto and herein incorporated by reference.

6. *Compliance with Laws.* In the performance of the services, university shall fully comply and require each of its subcontractors and assigns to fully comply with the requirements of any and all applicable laws, regulations, rule and orders of any governmental body having jurisdiction over the performance of this contract. University further agrees to the extent permitted by law to indemnify and hold company harmless from and against any costs, expenses, attorneys' fees, citation, fine, penalty and liability of every kind and nature which might be imposed by reason of any asserted or established violation of any such laws, order, rules and/or regulations.

Third Parties

1. *Other Agreements.* It is recognized that the work of some proposed project participants may be supported in whole or part under contracts and grants between the university and other parties. Prior to the participation of such an individual or the use of equipment owned by other sponsors in a joint study project, university will use its best efforts to identify and disclose to company any terms in those contracts which may conflict with company's obligations under this agreement. Company may, at its option, and in writing, agree to waive or

alter any such conflicting rights under this agreement in favor of those prior obligations so disclosed. If company does not agree to waive or alter any such conflicting right in a manner satisfactory to university, the individual will not be able to participate in that portion of the work under this agreement as to which such conflict arises. Company understands some university employees have private consulting agreements with third parties, to which the university is not privy and for which it disclaims all responsibility.

2. *Government Obligations.* Nothing in this agreement shall be construed to restrict the right of university to transfer to the U.S. Government such rights as the government may be entitled to under any agreement university may have, or may hereafter enter, with the government, whether or not consistent with the provision of this agreement.

3. *Exclusivity.* Funding for the research program will be exclusively by sponsor unless additional funds are made available to the university for its unrestricted use by the U.S. Government or private sources which are approved by sponsor. The university will consult with sponsor regarding the use of any equipment or facility in connection with the research program which has been acquired, in whole or in part, through U.S. Government funding. OMB Circular ____ is referenced herein as establishing the U.S. Government patent policy applicable to any government funding of the research program.

4. *Rights of Third Parties.* Company accepts that the university makes no representation nor gives any warranty that the use of information arising from the research undertaken under terms of this agreement will not infringe the rights of third parties.

5. *Interests of Prime Sponsor.* In the event that services to be performed under this agreement involve work covered under contracts which company may have as a government prime contractor or subcontractor, the obligations set forth shall be binding upon the university. Likewise, if these services to be performed include the receiving, handling or development of any government classified material, university agrees to com-

ply with all applicable security regulations and requirements and agrees to submit a confidential report to company immediately whenever for any cause university has any reason to believe that there is an active danger of espionage or sabotage affecting any work under such government contracts.

6. *Patent Associated Liability.* In consideration of company's financial and other support of the research project under ____ and of the patent work and cost thereof to be undertaken by company under this article ____, university and ____ agree they will make no claims against and hereby waive any claims they may have against company or its employees for injury, loss or damage resulting from acts of omission or commission by company, its employees or agents, in connection with the preparation, filing and prosecution of patent applications and the obtaining and maintaining of patents on university inventions. Each inventor, other than ____, of a university invention, no later than the time of signing a patent application thereon, shall be requested to agree, for the considerations recited in paragraph ____, to make no claims against and to waive any claims he may have against company or its employees for injury, loss, or damage resulting from acts of omission or commission by company, its employees or agents, in connection with the preparation, filing and prosecution of patent applications and the obtaining and maintaining of patents on university inventions. Should any inventor decline to so agree, any patent application on such university invention shall be filed and prosecuted and patents obtained and maintained by university, and the cost of such filing and prosecution and of obtaining and maintaining patents shall be borne by university.

7. *Related Research.* This paragraph pertains only to research by ____ and other persons assigned to, or working, on the research project. Company recognizes that such persons may from time to time receive financial support for research within the field of agreement from governmental or private nonprofit organizations and that it will generally be desirable to contin-

ue to receive some funding from such diverse noncommercial sources. However, support will not be accepted by university that would impair company's rights anticipated to flow from the research project, nor will support be accepted from any commercial party for research within the field of agreement. Before accepting support for research within or arguably related to the field of agreement, the research proposal shall be submitted to company for the purpose of mutual consultation on the questions of possible conflict and whether company would wish to be considered as the source of support. Such consultation shall be completed within _____ days, unless extended by agreement of the parties.

NAME
NON-DISCLOSURE AGREEMENT

1. Parties

1.1 For purposes of this Agreement, the term "Parties" shall collectively refer to (Name), a (entity type, state of origin) located at (address) (hereinafter "Name 1") and (Name), a (entity type) located at (address) (hereinafter "Name 2"). The term "Party" may refer to either (Name 1) or (Name 2).

1.2 The Parties hereby agree to perform their respective obligations as set forth within this Agreement subject to all applicable laws of any governmental body having jurisdiction over the performance of this Agreement.

1.3 The relationship between the Parties shall be that of independent contractors. Neither Party shall have authority to bind the other Party to any legal obligation.

2. Purpose of Agreement

2.1 The Parties acknowledge that the purpose of this Agreement is for (Name 1) and (Name 2) to exchange proprietary and confidential information so that (Name 1) and (Name 2) may evaluate the information and determine whether it may desirable to enter into a subsequent agreement with regard to said information.

3. Ownership of Products

3.1 Each Party shall retain all right, title, and interest in their respective Confidential Information disclosed under this Agreement.

3.2 Neither Party shall have any right to use of the Confidential Information other than for the purpose of this Agreement.

4. Term and Termination

4.1 The Agreement shall take effect upon the date of its acceptance through execution of all Parties.

4.2 The obligations under this Agreement shall be perpetual unless otherwise extended or terminated under the provisions of this Agreement.

4.3 Upon termination of this Agreement, each Party shall return to the other Party all Confidential Information that it has received under this Agreement.

5. Expenses

5.1 Each Party shall bear its own expenses in the performance of this Agreement.

6. Confidential Information

6.1 For purposes of this Agreement, the term "Confidential Information" shall refer to any idea, communication, prototype or work product, expressed or demonstrated in any manner, without requirement that it be reduced to a fixed medium, in which a Party may assert a proprietary claim of ownership.

x116.2 Confidential Information may be disclosed either orally, in a fixed medium, or by way of demonstration or enactment as follows:

(a) When disclosed orally, all information shall be treated as Confidential Information, unless the disclosing Party shall indicate otherwise in writing at the time of the disclosure;

(b) When disclosed in a fixed medium, the Confidential Information shall be identified and labeled "Confidential";

(c) When disclosed by way of demonstration or enactment, the Confidential Information shall be orally identified and described and, where possible, labeled "Confidential"; and

(d) Upon written request, each Party shall provide a signed, dated receipt which itemizes the Confidential Information transmitted under the terms of this Agreement.

6.3 Neither Party shall copy or reproduced any Confidential Information received from the other Party, except for such copies as are reasonably needed for the performance of this Agreement.

6.4 Each Party shall treat all Confidential Information as a trade are reasonably needed for the performance of this Agreement.

6.4 Each Party shall treat all Confidential Information as a trade

Appendix C

secret of the other Party and shall not disclose any Confidential Information to any other individual or organization unless and until expressly authorized to do so in writing by the disclosing Party. This obligation shall exist in perpetuity and survive the termination of this Agreement.

6.5 Each Party shall use all reasonable efforts to prevent the disclosures of Confidential Information to any other individual or organization.

6.6 Restrictions upon the disclosure of Confidential Information shall not apply to the following:

(a) information that was in the public domain at the time it was disclosed or falls within the public domain through no breach of this Agreement;
(b) information that was known to the receiving Party at the time of disclosure; or
(c) information that was disclosed after the written approval of the disclosing Party; or
(d) information that was made known to the receiving party from a source other than the disclosing Party without breach of this Agreement by any third Party.

6.7 In the event that either Party believes that any information should not be regarded as Confidential Information, the receiving Party shall give the disclosing Party written notice thereof within five business days following the disclosure, specifying the reason(s) for its position. If such notice is not given as provided herein, the information disclosed shall be treated as Confidential Information. If such notice is given and the disclosing Party does not accept the reason(s) stated, the information shall be treated as Confidential Information until such time as a determination is made under the dispute provisions of this Agremeent.

7. Attorney's Fees

7.1 In the event suit is brought, or arbitration initiated, or an attorney retained to enforce this Agreement, the prevailing party shall be entitled to recover, in addition to any other remedy, reimbursement for attorneys' fees, court costs, investigation costs and other related expenses incurred in connection therewith.

8. Arbitration

8.1   Any claim or controversy arising out of or relating to the performance of this Agreement, shall be settled by arbitration in (City), (State) in accordance with the Commercial Arbitration Rules of the American Arbitration Association, as herein amended. Judgment upon the arbitration award rendered by the arbitrator may be entered in any Court having jurisdiction thereof.

8.2   Limited civil discovery shall be permitted for the production of documents and taking of depositions. All discovery shall be governed by the Federal Rules of Civil Procedure. All issues regarding conformance with discovery requests shall be decided by the arbitrator.

8.3   Either Party may apply to any court having jurisdiction hereof and seek injunctive relief so as to maintain the status quo of the Parties until such time as the arbitration award is rendered or the controversy otherwise resolved.

9. Law and Forum

9.1   This Agreement shall be deemed entered into in the State of (State) and shall be construed in accordance with the laws of the State of (State) and of the United States. The parties stipulate that the proper forum, venue and court for any legal action taken with regard to this Agreement shall be held in the (Court) for (County), (State) or within a United States District Court for the State of (State).

10. Transfer of Obligations

10.1   Neither party may assign any of its duties or obligations without the prior written consent of the other party.

11. Entire Agreement

11.1   This Agreement supercedes all prior communications between the parties and constitutes the complete agreement with regard to those activities described herein.

12. Modification

12.1 This Agreement shall only be modified through use of a signed written document.

(Name)

_____   _____
Signature of Representative   Date

(Name)

_____   _____
Signature of Representative   Date

Appendix C

EXCLUSIVE LICENSE AGREEMENT
BETWEEN ——————— **AND** ———————

INDEX

Part I Implementation of Agreement

    1. Definitions 1
    2. Purpose of Agreement 2
    3. Transfer of Obligations 2
    4. Notices 2
    5. Entire Agreement 3
    6. Waiver 3
    7. Modification of Agreement 3
    8. Interpretation 3

Part II Relationship of Parties

    9. Parties 3

Part III Obligations of (Name 1)

    10. License 3
    11. Warranties 4
    12. Indemnification 4
    13. Cooperation 4
    14. Protection of Patent Rights 4
    15. Waiver of Claims 4
    16. Disclosure of U.S. Patent Application 4

Part IV Obligations of (Name 2)

    17. Ownership of Invention 5
    18. Cooperation 5
    19. License and Royalty Payments 5
    20. Reports 6
    21. License of Modifications 6
    22. Indemnification 7
    23. Protection of Patent Rights 7

Appendix C

Part V   Mutual Obligations

   24. Confidential Information. 7
   25. Conformation with Law 8

Part VI   Breach of Agreement

   26. Notice 9
   27. Right to Cure 9
   28. Force Majeure. 9
   29. Limitation of Remedies and Liability 9
   30. Dispute Resolution 9
   31. Law and Forum 10

PART 1
IMPLEMENTATION OF AGREEMENT

1. Definitions

1.1 For purposes of this Agreement, the following terms shall be defined as set forth below:

(a) "Parties" shall collectively refer to (Name 1) (hereinafter "Name 1"), a (State) (type of entity), located at (address) and (Name 2) (hereinafter "Name 2"), a (State) (type of entity), located at (address). The term "Party" may refer to either Name 1 or Name 2.

(b) "Invention" shall mean (narrative description) as described or claimed within U.S. Patent Application Serial No. (Number), together with Technical Information and any Modifications thereto which Name 1 may own or have the right to license during the term of this Agreement.

(c) "Modification" shall mean any substantial improvement, enhancement or other change in or to the Invention.

(d) "Patent Rights" shall mean U.S. Patent Application Serial No. Number, and all Modifications thereto by Name 1, and all other U.S. and foreign patents thereon which Name 1 owns or has the right to license during the term of this Agreement. In the case of an unissued patent, the scope of the Patent Rights shall be determined by reference to the claims that are asserted therein from time to time. In the case of an issued patent, the scope of the Patent Rights shall be determined only by reference to the issued claims.

(e) "Technical Information" shall mean any proprietary information with regard to the Invention which is not included within U.S. Patent Application Serial No. Number.

(f) "Product(s)" shall mean each of the elements set forth in Name 1's Patent Rights, the manufacture or sale of which by Name 2 would constitute, but for the existence of this Agreement, an infringement of Name 1's Patent Rights.

(g) "Confidential Information" shall mean any idea, communication, prototype or work product related to the creation, use, manufacture, distribution or sale of the Invention, expressed or demonstrated in any manner, without requirement that it be reduced to a fixed medium.

(h) "Gross Revenues" shall mean all revenues received from any sale of a Product, less charges for insurance, shipping, taxes, trade discounts, refunds, credits and returns, charges for replacement parts, and support services. In the case of a Product that is sold as part of a larger package, assembly or apparatus, Gross Revenues for that Product shall be in proportion to the ratio that the manufacturing cost of the Product bears to the manufacturing cost of the entire package, assembly or apparatus.

2. Purpose of Agreement

2.1   The Parties acknowledge that the purpose of this Agreement is to provide for the transfer by Name 1 of certain legal rights in the Invention to Name 2. The transfer shall allow Name 2 to engage in the manufacture and sale of Products.

3. Transfer of Obligations

3.1   Neither Party may assign any of its duties or obligations without the prior written consent of the other Party; except that Name 2 may assign this Agreement in its entirety to the purchaser of substantially all of the business of Name 2 to which this Agreement relates, provided that such purchaser agrees in writing to assume all the rights and obligations of Name 2 hereunder.

4. Notices

4.1   All notices under this Agreement shall be in writing. Service of notice shall be deemed adequate if (1) personally delivered, or (2) sent by certified mail.

4.2   Notices to Name 1 shall be sent to:

Name
Name 1
Address

4.3   Notices to Name 2 shall be sent to:
Name
Name 2
Address

4.4   Notices shall be sent to the addresses stated above, unless changed by written notice. Mailed notices shall be presumed received five (5) days after the notice is deposited in the United States mail.

5. Entire Agreement

5.1   This Agreement supercedes all prior communications between the parties and constitutes the complete agreement between Name 1 and Name 2 with regard to those subjects and activities described herein.

6. Waiver

6.1   No obligation or condition of this Agreement shall be waived by either party except by written consent. No act or failure to act shall serve as a waiver.

7. Modification of Agreement

7.1   No modification of this Agreement shall be binding upon either party unless in writing and signed by both Parties.

8. Interpretation

8.1   In the event that any provision of this Agreement is determined to be prohibited or unenforceable, the remainder of this Agreement shall continue in effect.

PART II.
RELATIONSHIP OF PARTIES

9. Parties

9.1   The relationship between the Parties shall be that of licensor and licensee. Neither Party shall have authority to bind the other Party to any legal obligation.

PART III.
OBLIGATIONS OF Name 1

10. License

10.1 Name 1 hereby grants to Name 2 the exclusive, worldwide right to manufacture and sell Products.

10.2 The initial term of the grant of rights shall be for a period of (time). In the event that a United States patent is issued to Name 1 for the Invention, then the term of the grant of rights shall be for the life of the patent.

10.3 Name 2 shall have the right to sublicense third parties, provided that any sublicense agreement satisfies the royalty provisions of Section 19 and otherwise conforms with Name 2 obligations under this Agreement. In the event of a sublicense, Name 2 shall give Name 1 notice of same and provide Name 1 with a copy of the sublicense agreement. Name 2 shall be responsible for collection of any sub-licensing fees.

10.4 In the event that Name 2 contests the validity of Name 1's Patent Rights, then Name 1 may immediately terminate the rights in the Invention granted to Name 2 as described in paragraph 10.1.

11. Warranties

11.1 Name 1 warrants that it holds all right, title and interest in the Invention and that to the best of its knowledge, the Invention does not infringe upon the rights of any third party.

12. Indemnification

12.1 Name 1 shall indemnify and hold Name 2 harmless from and against any claim by a third party that the manufacture or sale of a Product infringes upon the rights of said third party, provided that Name 2 shall give timely notification to Name 1 upon becoming aware of such a claim.

13. Cooperation

13.1 Name 1 shall cooperate in good faith with Name 2 to complete the performance of this Agreement.

14. Protection of Patent Rights

14.1 Name 1 may, but shall not be obligated to, institute legal proceedings against any third party for infringement of Name 1's Patent Rights. 14.2 Name 1 shall cooperate with Name 2 with regard to Name 2 instituting legal proceedings against any third party for infringement of Name 1's Patent Rights. Name 1 hereby consents to being named as a co-claimant in such legal proceedings.

15. Waiver of Claims

15.1 Upon execution of this Agreement and payment of all royalties accrued as of the date of this Agreement, Name 1 shall waive any claim that it may have against Name 2 or its officers, directors, employees or representatives based on actions prior to the date of this Agreement which relate to the Invention.

16. Disclosure of U.S. Patent Application

16.1 Upon execution of this Agreement, Name 1 shall provide Name 2 with a copy of Name 1's U.S. Patent Application Serial No. Number. Said copy shall be treated as Confidential Information.

16.2 Upon execution of this Agreement, Name 1 shall provide Name 2 with copies of all U.S. Patent Office actions and other communications with the U.S. Patent Office with regard to the Patent Rights. Name 1 shall also furnish to Name 2, for its review and comments, copies of all Name 1's responses and communications to the U.S. Patent Office prior to their submission.

PART IV.
OBLIGATIONS OF Name 2

17. Ownership of Invention

17.1 Name 2 hereby acknowledges that Name 1 is currently the sole owner of all right, title and interest in U.S. Patent Application Serial No. Number.

18. Cooperation

18.1 Name 2 shall cooperate in good faith with Name 1 to complete the performance of this Agreement.

Appendix C

19. License and Royalty Payments

19.1 Upon execution of this Agreement, Name 2 shall pay to Name 1 a non-refundable lump sum cash payment in the amount of (Number) Dollars. Said payment shall constitute a license fee and shall not be treated as an advance against royalties.

19.2 Name 2 shall also pay to Name 1 a royalty based upon the sale any Product. This obligation shall include any Product sold prior to the signing of this Agreement. The royalty shall be the higher of $(Number) on each individual Product or an amount equal to (Number %) Per Cent of Gross Revenues.

19.3 As a continuing condition of the grant of rights from Name 1 to Name 2, Name 2 shall pay to Name 1 minimum royalty payments of $(Number) per calendar year, which shall apply against the royalties described in paragraph 19.2. The minimum royalty payments shall be paid regardless of actual manufacture or sale of Products. The obligation to make minimum royalty payments shall commence in the calendar year beginning January 1, 1991. In the event that Name 2 fails to make minimum royalty payments in any one year, then all rights in the Invention granted by Name 1 to Name 2 shall terminate.

19.4 All royalty payments shall be made thirty days following each calendar quarter for royalties accrued within the quarter. The minimum royalty payment shall be made thirty days following each calendar year.

19.5 In the event that a determination is made that Name 1 does not hold all right, title and interest in the Invention, and as a result thereof Name 2 is obligated to pay royalties to a third party in order to continue to receive benefits from the Invention, then the royalty payments made by Name 2 to Name 1 shall be reduced in the amount of royalty payments made to the third party.

20. Reports

20.1 Name 2 shall maintain accurate records in accordance with generally accepted accounting principles. Said records shall be kept in sufficient detail so as to allow Name 1 to verify the accuracy of the royalty payments.

20.2 Name 2 shall prepare and present to Name 1 a report on Product royalties with each royalty payment.

20.3   Name 2 shall allow an audit of its records, with regard to royalties, by Name 1 during regular business hours and following reasonable notice. In the event that the audit of the records reveals a discrepancy of ten (10%) percent or more in Name 1's favor, then Name 2 will reimburse Name 1 for the cost of the audit.

20.4   Upon execution of this Agreement, Name 2 shall provide to Name 1 any Technical Information in its possession. Thereafter, Name 2 shall provide additional Technical Information as it becomes available.

20.5   All Technical Information and other documents provided by Name 2 under this Section 21 shall be treated as Confidential Information.

21.   License of Modifications

21.1   Name 2 hereby grants to Name 1 a non-exclusive, worldwide right to use any Modifications which are owned by Name 2. The use of the Modifications would be limited to research and development activities. The term of the grant would be for the longer of four years from the date of first sale of a Product, the life of any Name 1 patent in the Invention, or the life of any Name 2 patent in the Modification. Name 2 acknowledges that the rights received from Name 1 under this Agreement constitute full consideration for this grant.

21.2   In the event that Name 1 terminates the grant of rights in the Invention to Name 2, then Name 2 shall grant to Name 1 a non-exclusive, worldwide right to sublicense the manufacture and sale of any products which embody Modifications to the Invention. The term of the grant of rights would be for the longer of four years from the date of first sale of a Product, the life of any Name 1 Patent in the Invention, or the life of any Name 2' patent in the Modification. Name 1 shall pay to Name 2 a royalty which shall be the higher of $(Number) on each individual product or an amount equal to (Number %) Per Cent of Gross Revenues.

22.   Indemnification

22.1   Name 2 shall indemnify and hold Name 1 harmless against any claim of any third party with regard to any injury caused or related to the manufacture of and sale of Products, provided that Name 1 shall give timely notification to Name 2 upon becoming aware of such a claim.

22.2   In the event that Name 2 obtains any insurance policy that provides coverage with regard to the manufacturer and sale of Products, then Name 2 shall provide a copy of the policy to Name 1 and shall have Name 1, its officers, employees, named inventors and agents named as an additional insured within the policy.

23. Protection of Patent Rights

23.1   In the event that Name 1 declines to institute legal proceedings with regard to an infringement by a third party of Name 1's Patent Rights, as described in paragraph 14.2, then Name 2 may do so. Name 2 shall bear its own costs and shall hold Name 1 harmless as to all costs, judgments or awards not arising out of Name 1's own misconduct. Name 2 shall be solely entitled to any recovery, award or settlement received from the legal proceedings.

PART V:
MUTUAL OBLIGATIONS

24. Confidential Information

24.1   Confidential Information may be disclosed by either Party to the other Party, either orally, in a fixed medium, or by way of demonstration or enactment as follows:

(a) When disclosed orally, all information shall be treated as Confidential Information, unless the disclosing Party shall indicate otherwise in writing at the time of the disclosure;
(b) When disclosed in a fixed medium, the Confidential Information shall be identified and labeled ""'Confidential'";
(c) When disclosed by way of demonstration or enactment, the Confidential Information shall be orally identified and described and, where possible, labeled ""'Confidential"; and
(d) Upon written request, each Party shall provide a signed, dated receipt which itemizes the Confidential Information transmitted under the terms of this Agreement.

24.2 Each Party shall treat all Confidential Information as a trade secret of the disclosing Party and shall not disclose any Confidential Information to any other individual or organization unless and until expressly authorized to do so in writing by the disclosing Party. This obligation shall apply to Confidential Information disclosed prior to, concurrent with, or after the execution of this Agreement. This obligation shall exist in perpetuity and survive the termination of this Agreement.

24.3 Each Party shall use all reasonable efforts to prevent the disclosure of Confidential Information to any other individual or organization.

24.4 Restrictions upon the disclosure of Confidential Information shall not apply to the following:

(a) information that was in the public domain at the time it was disclosed or falls within the public domain through no breach of this Agreement;
(b) information that was known to receiving Party at the time of disclosure; or
(c) information that was disclosed after the written approval of disclosing Party; or
(d) information that was made known to the receiving Party from a source other than disclosing Party without breach of this Agreement by the receiving Party; or
(e) information reasonably required to commercialize the Invention through the manufacture and sale of the Product.

24.5 In the event that either Party believes that any information should not be regarded as Confidential Information, it shall give the disclosing Party written notice thereof within five business days following the disclosure, specifying the reason(s) for its position. If such notice is not given as provided herein, the information disclosed shall be treated as Confidential Information. If such notice is given and the disclosing Party does not accept the reason(s) stated, the information shall be treated as Confidential Information until such time as a determination is made under the dispute provisions of this Agreement.

24.6 Neither Party shall use Confidential Information of the other

Party, except for the performance of this Agreement and as otherwise specifically authorized in writing.

25. Conformation with Law

25.1 The Parties hereby agree to perform their respective obligations as set forth within this Agreement subject to all applicable laws of any governmental body having jurisdiction over the performance of this Agreement.

PART VI.
BREACH OF AGREEMENT

26. Notice

26.1 If either Party determines that the other Party has breached this Agreement, the non-defaulting Party shall give written notice of the breach to the other Party.

27. Right to Cure

27.1 The defaulting Party shall have thirty (30) from receipt of the notice in which to cure the breach. In the event that the defaulting Party fails to cure the breach, then the non-defaulting Party may terminate this Agreement, in addition to any other rights it may have.

28. Force Majeure

28.1 Neither Party shall be responsible for any delay in its performance of any of its obligations under this Agreement due to circumstances beyond its reasonable control.

29. Limitation of Remedies and Liability

29.1 In the event suit is brought, or arbitration initiated, or an attorney retained to collect any money due under this Agreement, or to collect a judgment for breach hereof, the prevailing party shall be entitled to recover, in addition to any other remedy, reimbursement for attorneys' fees, court costs, investigation costs and other related expenses incurred in connection therewith.

30. Dispute Resolution

30.1 Any claim or controversy arising out of or relating to the performance of this Agreement, shall be settled by arbitration in Denver, Colorado in accordance with the Commercial Arbitration Rules of the American Arbitration Association, as herein amended. Judgment upon the arbitration award rendered by the arbitrator may be entered in any Court having jurisdiction thereof.

30.2 Limited civil discovery shall be permitted for the production of documents and taking of depositions. All discovery shall be governed by the Federal Rules of Civil Procedure. All issues regarding conformation with discovery requests shall be decided by the arbitrator.

30.3 Either Party may apply to any court having jurisdiction hereof and seek injunctive relief so as to maintain the status quo of the Parties until such time as the arbitration award is rendered or the controversy otherwise resolved.

31. Law and Forum

31.1 This Agreement shall be deemed entered into in the State of Colorado and shall be construed in accordance with the laws of the State of Colorado and of the United States. The parties stipulate that the proper forum, venue and court for any legal action taken with regard to this Agreement shall be held in the District Court for Denver County, Colorado or within the United States District Court for the State of Colorado, subject to the requirement for arbitration set forth above.

WHEREFORE, the Parties agree to be bound by the terms and conditions of this Agreement which shall take effect as of the date of first sale of a Product, which is acknowledged to be the

_____ day of _____, 19____.

NAME 1

_____   _____
Signature of Representative   Date

Appendix C

NAME 2

_____   _____
Signature of Representative   Date

Index

A
antitrust, 152-153
arbitration, 154

B
bargaining ranges, 103
Bayh-Doyle Act, the, 14-15, 85
Berne convention, 160
"black box," 67
blue chip investments, 17
buyers
   financing, 133, 138, 139
   finding, 120, 122
   identification of, 49-50
   qualifying, 122
   technology and, 46, 48, 63
   valuation of technology, 51-53

C
calendars, 38-39
commercialization concept, 59

compensation, forms of, 143
Competitiveness Act of 1984, the, 72
computer-aided design (CAD) 48
computer simulations, 31
consulting firms, *See* facilitators
contract R&D transfers, 73-74
cooperative R&D agreements, 86-89
copyrights, 146, 148
   international, 160
corporate parent, 99
corporate profit system, 4
countertrade, 161
cycle model, 23

D

debt, 135
deep-pocket R&D, 10
demand-technical feasibility fusion, 22
development pipeline, 16
distribution channels. *See* marketing skills
due-diligence investigation, 107
Dun & Bradstreet, 122

E

economic proof, 60
equity, 136-137
European patent convention, 156
exclusive license agreement, 186-210
   liability and, 206-210
   patents and licenses in, 203-206
   period of agreement, 197-199
   recitals, 186-188
   research and reporting, 188-192
   sample, 216-229
   supervision, 192-197
   use of names in, 199-203
export restrictions, 160-161

Index

F
facilitators, 15
Federal Competitiveness Act, the, 153
Federal Laboratory Consortium, 70, 139
Federal Technology Transfer Act of 1986, the, 85
financial skills, 125-144
   accounting considerations and, 132
   buyer financing and, 133
   seller financing and, 132-133
   technology transfer costs and, 130-131
financing
   deal structure and, 140
   sources, 138
   types of, 133, 134-138
foreign exchange options, 162
Freedom of Information Act, the, 88-89
functional transfer model II, 65
funding sources, 136

G
gatekeepers, 104, 105
generic technology transfer, 7
global patenting, *See also* patents
global transfers, 5
go/no go analysis, 38 *see also* technology life cycle
government
   financing, 125, 127, 129, 139
   funding, 12-14
   technology transfers, state, 128
grant-backs, *See* compensation, forms of
graphs
   Bargaining Ranges, 103
   Before Writing the Plans: Factors To Consider In Constructing A Cooperative R&D Agreement, 86
   "Characteristics of Successful University-Industry Linkage, 15
   Commercial Concept, 59
   Comparison of Japanese and U.S. Technology by R&D Rates, 157

"Economic Impact of Innovations," 18
Four-Stage Process of Technical Advance, A, 3
Functional Transfer Model II, 65
Funding Sources and Expenditures, 136
Gatekeeper Network in a Geographically-Dispersed Organization, The, 105
"Industrial Outreached," 83
Innovation Process and Pace of Some Inventions, 47
Less Than Complete Sale, 67
Management Infrastructures in Academia, 85
Model for Linking Manufacturers with Technology Resources, 49
Offshore Legal Issues Reported in 1988, 151-152
Projected Transfer Time Line, 58
Projected Management Issues, 82
"R&D Success and Its Predicators," 5
Reverse Engineering Checklist, 78
Royalties Collected By Agencies: 1986-1988, 13
Science and Technology Incentives, 126
Science, Technology, and the Utilization of Their Products, Showing Communication Paths Among the Three Streams, 44
Scope and Performers, 137
Seller Checklist: Anticipating Buyer Needs and Questions, 54
State Government Technology Transfer Activity, 128
Straight Sale, 68
Strategies For Promoting Technology Transfer to the Private Sector, 102
Technical Advance Related to Demand-Technical Feasibility, 22
Technology Transfer Mechanisms, 66
Technology Transfer Situations, 100
Technology Transfer System (Participants), 8
"Top Transfer Performers," 10
Total Innovation Process, The, 28

H
hand-crafted production runs, 33
hedges, 96-97, 162

Index

I
independent contractors. *See* legal skills, business structures/relationships and
industrial outreach, 83
industry study, 60
innovation
  economic impact of, 18
  process, 28, 42, 47
intellectual property, 111, 145-146
international legal issues, 151
international transfers, 155-164
  global licensing, 156, 158
  protection, 156
inventions, American myth of, 4
inward licensing, 158

J
Japan, economic growth of, 163-164

L
legal rights, 46, 48-50, 79-80
legal skills
  business structures/relationships and, 150-157
  contracts and, 149
  dispute resolution and, 153-154
  fraud and, 149-150
  intellectual property and, 145-146
  licensing and, 147-148
  misrepresentation and, 149-150
  nondisclosure agreements and, 148-149
legislation, 73 *see also* individual Acts
liability, 206-210
license equity. *See* compensation, forms of
license transfers, 70-71
licensing agreements, 143-144
licensing fees. *See* compensation, forms of
life cycle of technology, 21-39

litigation, 154
loaned-servant transfers, 74

M
Madrid Agreement, the, 160
management strategies. *See* planning/management strategies
Manhattan Project, the, 118
marketing skills, 99-123
   bargaining ranges, 103
   benefits/needs, 99, 101
   end-users and, 101, 104
   incentives, 107-108
   national need for, 109
   Oak Ridge Approach and, 109-120
   presentation, 108-109
   private sector and, 102
   prospective buyers and, 104
market study, 60
mediation, 154
mergers, 74-76
minimum payment formulas, 61

N
National Technical Information Service (NTIS), 69-70
negotiation, 153-154
net present value (NPV), 141
niche marketing, 83
non-disclosure agreements, 211-215
NTIS. *See* National Technical Information Service
Nuclear Regulatory Commission, 161

O
Oak Ridge Approach, 109-120
Office of Technology Applications, (OTA), 111
one-time fees. *See* compensation, forms of
operating creativity, 97
organizational gatekeepers, 104-105
OTA. *See* Office of Technology Applications

Index

outright forwards, 162
outward licensing, 158
Paris Convention for the Protection of Industrial Property, 156, 160
partners. *See* legal skills, business structures/relationships and
Patent Cooperation Treaty, 156
patents, 146, 148, 223, 225
planning/management strategies
  capital, 92
  personnel, 93-94
  resources, 91-92
  tactics, 97
  technology, 94
  time, 94, 95
points of transfer, 39
price-fixing, 153
process technology, 150
product life cycle, 42
product technology, 150
product typing, 153
project management issues, 82
profit-sharing, 137
publication transfer, 69

R

R&D resources, predicators of, 5
resale transfers, 71-73
return on investment (ROI), 17
reverse engineering, 148
ROI. *See* return on investment royalties, 13, 137, 223
royalties, 13, 137, 223

S

sample agreements model, 75-85
science and technology incentives, 126
"secondary knowledge," 62-63
sellers
  buyer market and, 48
  checklist, 54

financing, 132-133, 138
identification of, 122
transfer technology, 62-63
single-application testing, 33
spin-off transfers, 76-77, 116
spot contracts, 162
start-ups, new business, 114
state government technology transfers, 128
step-wise refinement. *See* operating creativity
Stevenson-Wydler Act, 85
straight-line progression in technology development, 37-38
straight sale, 5, 68
sub-contracting, 66
swaps, 162

T

technology
  capitalists, 138
  catalysts, 122
  evolution of, 6-7
  incentives, 126
  insights, 122
  life cycle, 21-39
  pricing, 57
technology transfer
  alternatives, 9
  associations/organizations, 165-166
  books on, 168-170
  conferences/fairs, 167
  database/directories, 166-167
  newsletters/magazine/periodicals, 168
  training, 170
  see also technology transfer process
Technology Transfer Act, the, 69
technology transfer process, 41-63
  analysis, 45-46
  deal, 55
  economic proof and, 50-51

    inventory, 43-45
    legal control and, 46, 48-50
    negotiation, 54-55
    presentation, 53-54
    scientific proof, 50
    structure of, 55-56, 58
theft, 79-80
third-party broker, 72
third-party proofing sources, 51-52
time lines, 38-39, 58
trademarks, international, 159-160
trade secrets, 146, 148, 158-159

U
university-industry linkage, characteristics of, 15
U.S. Department of Commerce, 161
U.S. Department of Defense, 161

V
venture capital, 138

W
warranties, 206, 221
word-of-mouth marketing, 37
World Intellectual Property Organization, 160